Lecture Notes in Electrical Engineering

Volume 413

About this Series

"Lecture Notes in Electrical Engineering (LNEE)" is a book series which reports the latest research and developments in Electrical Engineering, namely:

- Communication, Networks, and Information Theory
- Computer Engineering
- Signal, Image, Speech and Information Processing
- Circuits and Systems
- Bioengineering

LNEE publishes authored monographs and contributed volumes which present cutting edge research information as well as new perspectives on classical fields, while maintaining Springer's high standards of academic excellence. Also considered for publication are lecture materials, proceedings, and other related materials of exceptionally high quality and interest. The subject matter should be original and timely, reporting the latest research and developments in all areas of electrical engineering.

The audience for the books in LNEE consists of advanced level students, researchers, and industry professionals working at the forefront of their fields. Much like Springer's other Lecture Notes series, LNEE will be distributed through Springer's print and electronic publishing channels.

More information about this series at http://www.springer.com/series/7818

Rinaldo Rinaldi · Romeo Bandinelli
Editors

Business Models and ICT Technologies for the Fashion Supply Chain

Proceedings of IT4Fashion 2016

Springer

Editors
Rinaldo Rinaldi
Dipartimento di Ingegneria Industriale
University of Florence
Florence
Italy

Romeo Bandinelli
Dipartimento di Ingegneria Industriale
University of Florence
Florence
Italy

ISSN 1876-1100 ISSN 1876-1119 (electronic)
Lecture Notes in Electrical Engineering
ISBN 978-3-319-48510-2 ISBN 978-3-319-48511-9 (eBook)
DOI 10.1007/978-3-319-48511-9

Library of Congress Control Number: 2016959254

Printed on acid-free paper

This Springer imprint is published by Springer Nature
The registered company is Springer International Publishing AG
The registered company address is: Gewerbestrasse 11, 6330 Cham, Switzerland

To our families that supported us every day.
To our Ph.D. students, young researchers
and colleagues that made this work possible.

Preface

The IT4Fashion is an industrial and scientific conference where fashion companies, brand, researchers, and software houses, from all the Europe, have the chance to discuss IT technologies as applied to the field of fashion and luxury.

The first edition was in 2011, and the event has constantly grown in terms of numbers, visibility, and importance for the operators in the sector.

Managing fashion products is challenging, not only for creatives and stylists, but also for supply chain managers. More and more, customers demand broad product variety and innovative products, and the competition in the fashion industry is ever more centered on the ability to react in a timely manner to changes in the customers' desires. Therefore, fashion companies have to balance the need to reduce lead times of collections, minimize stocks and obsolescence risks, acquire information from big data coming from digital and social media channels, guarantee a high level of quality, and get the customer involved in the processes of product development. Starting from a multidisciplinary approach, the IT4Fashion conference aims to collect and present new or improved supply chain business models and technologies applied to the fashion industry, to share knowledge among practitioners and researchers, and, in summary, to increase knowledge in the areas of Product Lifecycle Management (PLM) and Supply Chain Management (SCM) in the industry.

The 6th edition was organized over three full days—with plenary and parallel sessions. The first two days were dedicated to experts' keynotes and industrial case studies carried out by fashion companies and worldwide vendors. The third day was entirely dedicated to the scientific conference. During that day, researchers, practitioners, and students discussed the main challenges of IT solutions applied to the fashion industry.

This volume collect 24 selected papers presented during the scientific conference. The topics ranged from *product development* (design) to its *management after sale,* with strong attention to the new dynamics linked to the use of e-commerce and social media.

The conference dealt with the following themes:

- Fashion new product development
- Fashion design
- New materials for fashion products
- Product lifecycle management
- Wearable technologies
- Corporate social responsibility
- Business models and entrepreneurship
- E-commerce and omni-channel management
- Social and digital strategies for the fashion system
- Fashion operations and supply chains
- Challenges for the retail industry
- Brand management and strategy
- Mix of theoretical and practical education
- Fashion business, technology, and innovation
- Fashion marketing and communication strategy
- Business cases of fashion companies

Sincere thanks go to all authors, participants, and those who helped to organize and run the conference:

Hamideh Afsarmanesh, University of Amsterdam, The Netherlands
Gaetano Aiello, University of Florence, Italy
Kate Armstrong, University of Greenwich, London, UK
Liz Barnes, University of Manchester, UK
Verónica Arribas Barreras, ISEM Fashion Business School, Navarra, Spain
Alessandro Brun, Politecnico of Milan, Italy
Luis M. Camarinha-Matos, New University of Lisbon, Portugal
Fethi Çalişir, Istanbul Technical University, Turkey
Federico Caniato, Politecnico of Milan, Italy
Ricardo A. Cassel, Universidade Federal do Rio Grande do Sul, Brasil
Elisabetta Cianfanelli, University of Florence, Italy
Donatella Corti, University of Applied Sciences and Arts of Southern Switzerland (SUPSI), Switzerland
Pamela Danese, University of Padova, Italy
Stefano Dotti, University of Bergamo, Italy
Raffaele Donvito, University of Florence, Italy
Marcello Fera, Second University of Naples, Italy
Jonathan Gander, University of the Arts London, UK
Nevin Cigdem Gursoy, Istanbul Technical University, Turkey
Patsy Perry, University of Manchester, UK
Antonio Rizzi, University of Parma, Italy
Matteo Montecchi, The British School of Fashion, GCU London, UK
David Romero, Tecnológico de Monterrey, Mexico

Klaus-Dieter Thoben, BIBA—Bremer Institut für Produktion und Logistik GmbH, Germany
Quan Yu, Norwegian University of Science and Technology (NTNU), Norway
Sergio Terzi, Politecnico of Milan, Italy
Yi Wang, University of Manchester, UK

We hope this book serves as a starting point for reflection and discussion on highly innovative topics in the field of fashion and luxury, and we look forward to meeting you at the next edition of IT4Fashion in Florence, during April 2017 (www.it4fashion.org).

Florence, Italy
September 2016

Rinaldo Rinaldi
Romeo Bandinelli
Conference Chairs

Contents

Part I
Future Trends in the Fashion Industry

A New Research Agenda for Luxury Supply Chain Management?

Alessandro Brun, Federico Caniato and Antonella Moretto

Abstract The relevance of luxury industry is still increasing at double digit from an economic perspective at the global level; whereas this dramatic increase is not followed by a comparable improvement of research about Luxury Supply Chain Management. This paper aims at presenting some papers as well as some examples from key luxury case studies, with the purpose of addressing a new research agenda for Luxury Supply Chain Management. And Supply Management, Demand Management, Contract Management, Variety Management, and Sustainability are areas of research that are still under-explored, although the incredible potential high relevance for luxury Critical Success Factors.

Keywords Luxury fashion industry · Supply chain management · Demand management · Contract management · Supply management · Visibility · Inventory management · Sustainability

1 Introduction

The luxury industry has attracted the attention of researchers for many years, mainly focused on sociological, marketing and branding aspects (e.g. Vickers and Renand, 2003; Atwal and Williams, 2009). On the contrary little attention has been paid to the operations and SCM perspective. But the luxury industry today plays a key role in the economy of several countries, due to various phenomena such as the growth of the affluent consumer segment in key countries such as China, the

A. Brun (✉) · F. Caniato · A. Moretto
Department of Management, Economics, and Industrial Engineering, Politecnico Di Milano, Milan, Italy
e-mail: alessandro.brun@polimi.it

F. Caniato
e-mail: federico.caniato@polimi.it

A. Moretto
e-mail: antonella.moretto@polimi.it

R. Rinaldi and R. Bandinelli (eds.), *Business Models and ICT Technologies for the Fashion Supply Chain*, Lecture Notes in Electrical Engineering 413,
DOI 10.1007/978-3-319-48511-9_1

increasing sophistication of consumers all over the world, and the ability of traditional players to distribute and sell their product on a global scale. For this reason, Supply Chain Management has increasingly become relevant also in this specific context, since luxury firms are now becoming large, international players who need to face complex challenges in developing, sourcing, producing and distributing their products, often on a global scale.

The seminal paper by the Politecnico di Milano research team (Brun et al., 2008; Brun and Castelli, 2008) on Logistics and Supply Chain Management in the high-end, luxury segment of the fashion industry were published in 2008. A research on Scopus indexed journals shows that—as of today—only 90 papers appear to have both the words "Luxury" and "Supply Chain" in the title or abstract or keywords, and only a fraction of them are actually pertaining to the Supply Chain Management of luxury goods domain.

By reviewing the existing literature, we can observe that several aspects of Supply Chain Management (SCM) in the luxury industry are indeed emerging as highly relevant and worth a deeper investigation. For example, Nueno and Quelch (1998) addresses that the whole supply chain is relevant for luxury companies; in the same vein Caniato et al. (2009) and Caniato et al. (2011) have investigated supply chain strategies of luxury industries, showing how consolidated models do not fit very well this specific context and proposing a new taxonomy. Macchion et al. (2015) focused on supply network strategies, analysing how they are dynamically managed by fashion luxury firms to align with evolving marketing strategies. Luzzini and Ronchi (2010) have analysed the organization and practices of purchasing departments of luxury companies, showing several peculiar characteristics and the need for further development. Brun and Moretto (2014) showed how quality is managed by luxury firms, since this is a paramount performance in this context. Caniato et al. (2012) have shown that the topic of supply chain sustainability is highly relevant for fashion companies, including luxury ones, with specific implications to be addressed. Brun and Moretto (2012) analysed contract design in the luxury jewellery industry as a tool to improve supply chain performance. New product development is also a key topic for luxury companies, needing a specific consideration, as shown by Caniato et al. (2013) and Caniato et al. (2014). This is particularly true in the context of globalisation, which interplays with outsourcing decisions, in both new product development, sourcing and production processes (Caniato et al., 2015; Macchion et al., 2015) as well as in the broader supply chain innovation perspective (Caniato et al., 2013).

This brief overview confirmed that every aspect of Supply Chain Management, including both strategic and operational ones, is not only relevant for luxury firms, but is also worthwhile deeper investigation, since very often the models and methods developed for other industries do not apply well or do not provide good performance when transferred to the luxury context.

Moreover, an extensive analysis of literature on Luxury and Luxury Management allowed us to single out the following list of Luxury Critical Success Factors (CSFs) (for detailed list of reference, please refer to Brun and Castelli, 2013):

- Consistently delivering ***premium quality*** in all the products in the line and along the whole Supply Chain, both through superior material quality and conformity to product specifications;
- A ***heritage of craftsmanship***, which ensures the necessary expertise for manufacturing high quality objects;
- ***Exclusivity*** obtained through the use of naturally scarce materials, limited editions, limited production runs, selective distribution and the creation of waiting lists;
- A marketing approach that combines product excellence with ***emotional appeal***; for instance, an appealing product display provides customers with an enhanced shopping experience, and the atmosphere at the point of sale reflects the values associated with the brand;
- The ***global reputation*** of the brand, which conveys the idea of world-class excellence;
- A ***recognisable style and design***, which means that consumers don't need to see the label to recognise the brand in some cases. For the luxury goods market, tangible features are insufficient. Customers must also respond to the product emotionally due to the product design and aesthetic;
- An association with a ***country of origin*** that has an especially strong reputation as a source of excellence for a certain product category, such as Champagne from France;
- Elements that establish ***uniqueness***, such as minor imperfections in hand-blown crystal vases;
- Superior ***technical performance*** for brands based on technical expertise, such as sports cars. Best-in-class technical performance appeals to customers emotionally and allows them to distinguish luxury products from ordinary ones. For this product feature, continuous innovation can sustain product positioning;
- The ***creation of a lifestyle*** that allows the customer to share in a unique lifestyle, which can be recreated in everyday life by possessing the luxury product;
- An ***ethical and sustainable*** Supply Chain. Focusing on such practices as responsible use of scarce resources, ethical work conditions, fair agreements with suppliers, in terms of implementation of good practices and consequent communication of endeavours and results.

The analysis proved that companies could pursue a luxury positioning for their brands and products (and apply the appropriate premium price) by implementing processes allowing them to strengthen one or more of said CSFs.

This finding simplifies a lot the definition of a future research agenda: there are still many unexplored areas in Luxury Supply Chain Management, and all of them will be analysed employing the same perspective of the impact of current and future practices on Luxury CSFs.

Therefore, the goal of the present paper is to provide an overview of the most relevant aspects of Supply Chain Management for luxury firms, highlighting the most relevant challenges for managers in this context, which are also the most interesting directions for future research.

We will leverage on several case studies performed over an extensive time span to discuss how leading luxury firms have evolved their Supply Chain Management, thus showing the emerging best practices, as well as the challenges that are still open for both managers and researchers.

2 Research Approach

This paper is developed based on several case studies developed by the Politecnico di Milano research team since 2006. During a decade of research, around 150 fashion and luxury companies were analysed through case studies, action research projects and surveys. Our focus is on the worldwide renowned Italian fashion luxury products, whose brands have achieved international recognition and are facing in the last years the challenge of developing Supply Chain strategies that can support them in competing globally. For this reason, the industries considered are those related to the traditional Italian products as well as the one consistent with the definition of luxury, such as, e.g., shoes, apparel, accessories. Interviews were conducted both in Small-Medium Enterprises (SME), which usually face great challenges in building global Supply Chains, and in large firms, who were SMEs once and have now grown to support their market expansion. The choice of such a varied sample—as regard company size—makes it representative of the actual composition of the Italian fashion luxury market, which is made of very few fashion giants with billionaire turnover (i.e. Gucci, Armani, Prada), dozens of—equally renowned—medium-sized firms (i.e. Tod's, Versace) and plenty of smaller players, some of which owning worldwide known luxury brands (i.e. Iceberg, Liu Jo, Aspesi, Marinella). In several cases, pivotal companies of the industry were also interviewed multiple times over the years, thereby performing an in depth longitudinal analysis of the cases as well as addressing the main challenges the industry has faced from a Supply Chain perspective. This paper has a twofold objective: on the one hand, it aims at identifying the main challenges for the industry as well as the main critical areas where companies should be excellent to overcome the competition and be successful; on the other hand, the paper aims at proposing some key areas of research for scholars, for providing some insights about hot topics as well as interesting research questions also from an academic point of view.

3 Discussion of the Research Agenda

3.1 Supply Management

In the area of purchasing and supply management, luxury companies are used to have a very high number of suppliers, often small and highly specialized, each of them perceived as highly strategic. Suppliers are directly involved in key processes such as the development of the new collections of products, the "engineering" of the designers' ideas into a product, the search for new materials and techniques to transform creative ideas into physical products. They are usually highly diversified in many respects: most are very small, local firms, while others are large multinationals; some are specialized in niche materials or operations, often unique and very difficult to replace, while others provide more standard materials and activities; some work exclusively for one customer, others have several customers, sometimes from different industries.

Let's consider the case of Company A, a major player in the luxury fashion industry worldwide, with a strong Italian identity, still preserved despite the acquisition by an international group. The company has historically outsourced its production to a large network (a few hundreds of medium, small and very small contractors, including both first and second tier, all located in Italy and most of them close to the company headquarters). These contractors work almost exclusively for the company and are closely managed and controlled, they are often family-owned firms, highly specialized, with highly skilled workforce. Company A is in charge of product development, but all production activities are performed externally. Company A takes care of planning production activities and sourcing raw materials and components to supply the contractors. Raw materials (such as tanned leather and fabrics) and components (such as metal parts and trimmings) come from Italy as well as much as possible, with some force exceptions (e.g. exotic skins such as snake and crocodile are not available locally). Also with these suppliers there is a long lasting collaboration and a close relationship and they are often responsible of superior ***technical performance***.

Until recently, the management of this broad and complex supply base was quite unstructured: contractors and suppliers were considered as a key strategic resource of the company and were treated very well: they were guaranteed orders all around the year, they were offered good prices and very fast payment terms, they were monitored almost only in terms of product quality, and in general they were all considered equally important and treated as such.

In the last few years however several new challenges arose: the economic and financial crisis affected also Company A's Supply Chain, with a more unpredictable demand of final products, more demanding targets by the group headquarters, more difficult access to credit in particular for medium and small suppliers, a stronger attention to sustainability issues by both the customers and other stakeholders such as NGOs and regulators. All these challenges requested a radical revision of the supply management strategy and practice.

First of all, Company A realized even better how much it depends from its supply base and therefore decided to both rationalize it, by selecting the best suppliers, and to provide even more support to them. This strategy was deployed through a comprehensive set of actions. First of all, the creation of a systematic and extensive vendor rating system, covering a broad set of performance (including quality, delivery, flexibility, cost, financial strength, technological competence, compliance and sustainability). Second, the set up of an agreement with a major financial institution to provide easier and cheaper access to finance for the best suppliers, exploiting their being strategic suppliers of Company A (which has a very solid financial position). Third, Company A even decided to revise its strategy of total outsourcing, starting to invest directly in the equity of some key suppliers (both contractors and suppliers of raw materials), to extend its control on key actors.

This case provides several interesting insights on the role of supply management for luxury fashion companies, and some key directions for development of both research and practice in this field.

This case shows how supply management in this industry has been underestimated, if not neglected, until recently, but now many companies, starting from the leading ones, have deployed significant efforts to revise and improve their management of the supply base, realizing how critical and strategic it is. In fact, supply base has controlled a big portion of value, thereby strongly impacting on ***exclusiveness***, ***uniqueness***, and Global ***brand reputation***. In doing so, they have started adopting traditional tools and techniques, such as spend analysis, portfolio management, vendor rating, supplier development, etc. Such tools and techniques have proved to be applicable also in this context, however, given the peculiarities of the industry, they often lead to different suggestions and conclusions. For example, a luxury brand cannot help working with very small suppliers, when they are a unique asset, providing exclusive competencies, therefore they need to be maintained and protected, while at the same time supported in their development. Also outsourcing starts to be questioned in this peculiar context, with major players starting to insource key competencies: this choice would strongly impact on the ***country of origin*** effect as well as ***heritage of craftsmanship***, often controlled by suppliers. While in other industries large companies tend to stretch payment terms to reduce their working capital, we increasingly see luxury firms provide financial support to their strategic suppliers, also adopting innovative Supply Chain Finance solutions, for fostering an ***ethical and sustainable Supply Chain***. For example, Company B is a key Italian company that decided to implement a structured Supply Chain Finance approach with the weakest links of its Supply Chain, for assuring the survival of these companies from a financial point-of-view; the relevance of this topic is confirmed by the big project jointly developed by Sistema Moda Italia and Unicredit, aiming at supporting the whole industry and value chain with a big Supply Chain Finance project.

This evidence suggests interesting research directions on supply management for luxury firms, aiming to investigate and develop innovative models and tools to address the specific characteristics of this peculiar context. At the same time, there are very important implications for managers as well, who need to innovate the way the supply base is managed in luxury firms, not just adopting the methods developed in other industries, but developing more specific solutions. Traditional portfolio management approaches may not be applied in the same way they are used in other industries, but the development of appropriate methods would help companies in reducing the number of suppliers and increasing the level of control and effectiveness in supply management, by implementing partnership just with really key and strategic suppliers.

3.2 Data Management and Visibility

Luxury fashion companies usually do not forecast demand for the high impact of totally new products in each collection. Five years ago, we interviewed Company C, an underwear company mentioning almost 100% of companies totally revised one collection after another, without commonalities in terms of either materials or components: for this reason, the company preferred using either a make-to-order or a purchase-to-order approach. As a matter of fact, this approach reads very consistent with the main CSFs of luxury—such as ***exclusiveness*** and ***uniqueness***–but not pursuable in the long-term anymore from a Supply Chain point of view. The company was required to revise its Supply Chain approach in this direction, for assuring a consistent management of product portfolio.

Whereas, early results from our research show that actually just a part of the seasonal collection is totally new although most of the collection is based on either carry over products or products that are an incremental evolution of existing ones, often using the same materials and components. For example, Company D, a key accessory Italian company, leader in leather good products addressed that 40% of the total collection is covered by carry-over products whereas the remaining 60% pertains to new products. Nevertheless, an additional 20–30% of those 60% is realised on existing materials or new products based on previous models or colours.

In this situation, the possibility of implementing a demand management approach would be strongly simplified. By developing appropriate demand management and forecasting methods, e.g. exploiting both historical data and market insights to better forecast also new products, as well as developing more collaborative approaches with retail, would increase the visibility along the Supply Chain with an expected reduction of both out of stock as well as overstock and the consequent need to sell products in outlets at a discounted price. This approach is particularly pursuable in an industry where most of the companies are totally controlling the retail chain: in the last years, luxury companies are in fact reinforcing their retail network, eliminating as far as possible both multibrand stores and franchising stores. This new way of managing demand management is fundamental for luxury CSFs, in particular ***exclusivity*** and ***uniqueness***: avoiding an excess of demand to sell at a discounted price would not negatively impact on exclusivity and uniqueness, by maintaining the premium price consistent with ***premium quality*** and the high positioning, typical of luxury companies.

Moreover, a new phenomenon for luxury companies is the "See Now Buy Now": presented by companies such as Burberry, Peuterey, Tom Ford, and Prada: this new management of the seasonal collection development implies that products are available for sales into the stores immediately after catwalks. As a matter of fact, in this new model companies cannot pursue either a make-to-order or a purchase-to-order approach anymore, with an upside down of traditional model. In this new competitive arena, traditional Supply Chain measures such as service level and timeliness would become more and more relevant, thus making a good Demand Management process a key element for success. About this new phenomenon,

Giorgio Armani addressed: « *For making actual this revolution, we would need to operate along the whole Supply Chain, to make the change actual and effective and avoiding that it is just another communication lever. As a matter of fact, this would require time but especially a key strategy, to foster and make operative at all the Supply Chain levels*". The management of this new approach, with a total revision of traditional deadline, would impact on the company's capabilities of controlling key CSFs, such as ***creation of a lifestyle*** and ***emotional appeal***.

3.3 Contract Management

To the best of our knowledge, Brun and Moretto (2012) is the only study published so far analysing contract design as a coordination mechanism between Brand Owners (BO) and Distributors/Retailers of luxury goods. Yet arguably, also considering results in other industries, contracts can be a very powerful coordination mechanism between two actors in the Supply Chain.

Nonetheless, the up-close analysis of a wide number of distribution agreements for such personal luxury goods as watches and jewellery, and perfumes and cosmetics, signed between brand owners and distributors, department stores, or single independent retailers, provided evidences that the potential of contracts is far by being exploited by Brand Owners when dealing with independent trade and distributors. On the contrary, analysed contracts are often afflicted by some relevant weaknesses:

- **no incentives for holding the "right" amount of stock.** The well-known newsvendor model clearly demonstrates the intuitive result that retailers should overstock (keep a stock larger than the expected demand) whenever the stock-out cost is higher than the cost of overstock (per unit of product sold/unsold), and understock in the opposite case. Brand Owners should make sure that the ratio Overstock cost versus Understock cost be the same for retailers and for the whole Supply Chain (be it through buy-back clauses or other incentive mechanisms). A wrong amount of stock might imply unsold products and stock management policy that might have a negative on ***exclusivity***, ***emotional appeal***, ***brand reputation*** and ***uniqueness***.
- **no incentives for anticipating forecast information.** Quite frequently, especially in far-away markets, retailers have a better knowledge of their customers and can have good insights over the latest trends in buying behaviours. Brand Owners could have great benefit, if they could receive accurate forecast information in advance from retailers. Yet such clauses as incentives (e.g. extra discounts) linked to forecast accuracy, revenue-sharing mechanisms, let alone using forecast information as a (partially) binding commitment for future orders, are seldom used. This lack of information would potentially deteriorate several CSFs, such as ***exclusivity***, ***emotional appeal***, ***uniqueness*** and ***creation of a lifestyle***.

- **no real time visibility of in-store stock/sell-out data.** This is a hugely debated topic also in other industries. Yet in the luxury industry visibility of in-store stock and sell-out data could really make a difference. What should a BO do if, in a country where it doesn't have Directly Operated Stores (DOS), a top client is looking for a crocodile bag—which allegedly should be in stock in 3 authorised resellers (with 1 piece each)? This wrong coordination of data as well as a wrong management of different countries would directly impact on ***brand reputation.***
- **consignment (goods are in the store but still owned by the BO; the stores pays the BO for the goods only when they are sold).** When working with consignment stock, the retailer has no incentives to make good forecasting. Moreover, unless the retailer works with all brands in consignment (which is the case, e.g., of some "iconic" boutiques or concept stores), the brands in consignments could be regarded as step-children in case of undecided customers (i.e. the sales-rep prefers to push the sale of an item in the "owned stock" side of the shelf, and leave the consigned stock unsold). This approach would directly impact on ***brand reputation*** of brands managed through consignment stock.
- **Product assortment width and depth decided by the BO.** There are cases in which the store manager (or buyer) knows better which items are going to be slow movers. In many instances, forcing the store to keep them in stock is questionable a strategy. But without a sharing of responsibility in defining and managing product portfolio might deteriorate ***exclusivity***, ***emotional appeal*** and ***brand reputation***.

Early results from our research show that with a better-designed contract, the expected profitability for the Supply Chain as a whole could soar. Therefore, the development of innovative contractual agreements, tailored to the specificity of the luxury industry, may provide a strategic benefit to the whole industry.

3.4 Variety Management

The impact of product variety on SCM complexity, and strategies and practical approaches to manage and reduce variety, are well explored topics in operations management, and plenty of literature exists, discussing the trade-off between the costs and benefits of variety, both in product design and manufacturing. In the classic "Variety Reduction Programme" (VRP) approach, when employing specific components and materials variety costs increase; conversely, the functional cost reduces because the best-suited components are used in each specific application. Dually, reducing the number of components and materials reduces variety costs, but better performing components are used unnecessarily to fulfil some assembly responsibilities, thus increasing the functional cost (Suzue and Kohdate, 1990).

This trade-off has been deeply analysed and it can be considered as a well-acknowledged management issue. Nevertheless, while defining the optimal amount of variety, other elements intervene, more related to organisational or human factors, which are often neglected.

In luxury-fashion, when launching new collections, variety (of both materials AND end-products) seems always to be positioned on the right-end-side of the theoretically optimal point. Clearly enough, extreme variety is used to sustain the luxury positioning, and some items may even be developed without the real goal of being sold (e.g. boots or a belt to complete an outfit; an outrageously expensive watch to show our ability to master the ultimate complication); yet this careless approach to variety skyrocketing does create some problems, such as dealing with minimum order quantities in e.g. fabrics, how to handle unsold items with damages on ***exclusivity*** and ***uniqueness***, impossibility to hold the whole collection in smaller stores with problems of ***brand reputation*** and ***emotional appeal***, etc.

If, from the one hand, the luxury fashion is not yet ready to implement Variety Reduction Programmes, for sure some basic actions (such as keeping the ratio between seasonal and carryover SKU within a certain range; increasing the communality of some raw materials and components; creating a differentiated decoupling point depending on communality of items and coefficient of variation of demand forecasting error) could help a better and healthier inventory management at all levels of the Supply Chain. This approach needs to be developed with a focus on the luxury critical success factors, in order to improve the Supply Chain performance while preserving the competitive advantages.

3.5 Sustainability Management

For several years, luxury and sustainable read as an oxymoron, but today literature as well companies are demonstrating a stronger attention towards these issues. Burberry clearly mentioned this year that luxury cannot exist anymore without sustainability; Pinault addressed sustainability as the key asset of the overall Kering Group: "*Sustainability is an opportunity. It creates long-term value for stakeholders and it is optimistic, not constricting*", said last year. But for making companies more sustainable, luxury companies have to totally and deeply revise their strategy, operating both at the product and at the Supply Chain level. For example, Company E, an Italian company part of a key French group has started the implementation of sustainable practices since several years through the implementation a Life Cycle Assessment for understanding where the major environmental impact is; then the company implemented a revision of Supply Chain processes (e.g., green vendor rating, supplier collaboration for sustainability, revision of logistics flows, etc.) and finally a revision of the main procedures for seasonal collection development (e.g., adoption of organic cotton).

Nevertheless, the implementation at the company level might be not sufficient. As a matter of fact, in the luxury industry the brand owner has the direct control just on a small part of the overall value generated along the Supply Chain: for this reason, it is not sufficient to implement practices at the company level otherwise the impact would be minimal and not really impacting. The real challenge is the extension of very advanced practices along the whole Supply Chain, by involving first- and second-tier suppliers, as well as distributors and retailers. Focal company is in the key position for driving this change and for fostering a real change, which is not pursued just for marketing and communication reasons.

One question arises: why pursuing sustainability? Is it just a trendy word or a real value is underlined? Quoting Pinault once more, "*Sustainability is not a constraint, it's a field of opportunities at every level. It creates value in multiple ways–efficiency, innovations in processes and product development*". Companies address that they implemented these procedures because a revision of Supply Chain design and management with a sustainable purpose would increase both external performance—***brand image***, customer loyalty, brand awareness—as well as internal performance at the Supply Chain level—***superior product quality***, process efficiency; moreover, sustainability is a key CSF for achieving an ***ethical supply chain*** but it is also perceived as one of the strongest and more relevant lever for increasing innovation, a key and fundamental CSF for luxury company, through the achievement of ***superior technical performance***.

The lesson learned is that sustainability is not really a choice anymore: companies should decide whether they want to drive the change or just becoming compliant with regulation thereby following choices of either governments or competitors. But this is the current trend: it is just a matter of developing an approach to sustainability that is aligned with the specific characteristics of the luxury industry, in order to preserve its peculiar features.

4 Conclusions and Future Developments

This paper started by a structured literature review about Supply Chain Management and luxury industry, by demonstrating still an embryonic level of advancement of this stream of literature. The first attempts are identified in 2008 but in the last 8 years no more than 90 papers can be listed about this topic. Nevertheless, case studies would suggest a strong need of research in this area. Examples of research questions that researchers can ask are the following:

- Which is the right management approach for coping with a distributed supply network? Which is the impact of Supply Chain Finance solution in the management of key supply base?
- Which are the right tools for demand forecasting for luxury companies? Which is the impact of the new phenomenon of "See Now Buy Now" on the whole Supply Chain Management?
- How much can the profitability of the whole Supply Chain increase thanks to a consistent and well design contract management approach?

- Is Variety Programme Management pursuable also for luxury companies? Which are the impacts of these programs for traditional Supply Chain performance as well as for CSFs?
- Which is the real value of sustainability for luxury companies? Which is the most effective way for implementing sustainable along the whole Supply Chain?
- And several others…

Upmost critical points pertain to the incredibly high impact that said Supply Chain Management approaches would have on luxury CSFs. It is not a matter of efficiency at all, but these are strategic levers for strongly increasing both the value and the competitiveness of the industry. As a matter of fact, the cases reported above would address the incredible potential impact on Supply Chain Management choices on CSFs: Supply Chain Management is not just a tactical or operational choice anymore, because a wrong management of those processes would irreparably impact on CSFs, with important damages.

By concluding, this paper is a preliminary investigation of a possible research agenda for the next few years of the industry: this list might be valuable for research, thereby supporting academics in identifying new areas and new streams of investigation but this list can also support managers is addressing the main challenges of the next future. Because the success of tomorrow is build up on Supply Chain Management processes and successful companies have already found out.

References

Atwal G, Williams A (2009) Luxury brand marketing–the experience is everything! J Brand Manage 16(5/6):338–346

Brun A, Caniato F, Caridi M, Castelli C, Miragliotta G, Ronchi S, Sianesi A, Spina G (2008) Logistics and supply chain management in luxury fashion retail: empirical investigation of italian firms. Int J Prod Econ 114(2):554–570

Brun A, Castelli C (2008) Supply chain strategy in the fashion industry: developing a portfolio model depending on product, retail channel and brand. Int J Prod Econ 116(2):169–181

Brun A, Castelli C (2013) The nature of luxury: a consumer perspective. Int J Retail Distrib Manage 41(11/12):823–847

Brun A, Moretto A (2012) Contract design and supply chain management in the luxury jewellery industry. Int J Retail Distrib Manage 40(8):607–628

Brun A, Moretto A (2014) Organisation and supply chain for quality control in luxury companies. J Fash Mark Manage 18(2):206–230

Caniato F, Caridi M, Castelli C, Golini R (2009) A contingency approach for SC strategy in the Italian luxury industry: do consolidated models fit? Int J Prod Econ 120(1):176–189

Caniato F, Caridi M, Castelli C, Golini R (2011) Supply chain management in the luxury industry: a first classification of companies and their strategies. Int J Prod Econ 133(2):622–633

Caniato F, Caridi M, Crippa L, Moretto A (2012) Environmental sustainability in fashion supply chains: an exploratory case based research. Int J Prod Econ 135:659–670

Caniato F, Caridi M, Moretto A (2013) Dynamic capabilities for fashion-luxury supply chain innovation. Int J Retail Distrib Manage 41(11/12):940–960

Caniato F, Caridi M, Moretto A, Sianesi A, Spina G (2014) Integrating international fashion retail into new product development. Int J Prod Econ 147:294–306

Caniato F, Crippa L, Pero M, Sianesi A, Spina G (2015) Internationalisation and outsourcing of operations and product development in the fashion industry. Prod Planning Control 26(9):706–722

Luzzini D, Ronchi S (2010) Purchasing management in the luxury industry: organization and practices. Oper Manage Res 3(1–2):7

Macchion L, Moretto A, Caniato F, Caridi M, Danese P, Vinelli A (2015) Production and supply network strategies within the fashion industry. Int J Prod Econ 105, 173–188

Nueno JL, Quelch JA (1998) The mass marketing of luxury. Business Horizons, November/December 1998

Suzue T, Kohdate A (1990) Variety reduction program: a production strategy for product diversification. Productivity Press, Massachusetts, Cambridge

Vickers JS, Renand F (2003) The marketing of luxury goods: an exploratory study. Mark Rev 3(4):459–478

Author Biographies

Alessandro Brun Prof. Alessandro Brun is an associate professor of quality management at Politecnico di Milano. His research interests cover application of Supply Chain Management, Quality Management and Operations Improvement methodologies in high value added sectors, with special focus on Luxury Management. With more than 100 papers published on scientific journals and international conferences on the above topics, is one of the most prolific authors in the worldwide scientific community on the specific "Luxury Supply Chain". In 2012, he created and launched the "International Workshop on Luxury Retail, Operations and Supply Chain Management" - the first international conference on Luxury Supply Chain.

Federico Caniato Federico Caniato is Associate Professor at the School of Management of Politecnico di Milano, teaching Supply Chain and Purchasing Management in both undergraduate and graduate courses.

At MIP Politecnico di Milano Graduate School of Business, he is the director of the Master in Supply Chain and Purchasing Management. His research interests are in the fields of Supply Chain and Purchasing Management, in the last years he has focused on Sustainability and Supply Chain Finance. He is the director of the Supply Chain Finance Observatory, the leading research initiative in Italy on the topic, in close collaboration with the international Supply Chain Finance Community. He authored several international publications on various Operations Management journals, and he is Associate Editor of the Journal of Purchasing and Supply Management.

Antonella Moretto Antonella Moretto is Assistant Professor at Department of Management, Economics and Industrial Engineering of Politecnico di Milano. She got a PhD at Politecnico di Milano with a thesis about the internationalization of Italian fashion companies in emerging market. During the period 2013–2015, she collaborated as consultant in the area of Operations and Supply Chain Management. Her main research interests are related to purchasing and supply management, sustainable supply chain management and global supply chain management.

Publication Trends in Supply Chain Management in the Fashion Industry

Basak Cetinguc, Eyup Calik and Fethi Calisir

Abstract The main objective of this research is to conduct a scientometric analysis of supply chain management in fashion- industry literature. The data were gathered from Web of Science database. "Fashion Industry" and "Supply Chain Management" were used as keywords to perform this research. Without narrowing down the time bounds, this search identified 123 publications related to supply chain management in the fashion industry were published up to that date. Though there have been annual ups and downs in the number of publications, the annual number of publications increased from one in 2000 to 18 in 2015. The most productive year was 2014 with 23 papers. Articles were the most popular documentation style with 94 papers, and the other documentation types were conferences, reviews, patents and editorial papers, in descending order. Moreover, 91% of publications were written in English, and China was found to be the most productive country. Concerning research areas, there was a wide range of fields, while the most trendy research area was engineering with 26%. Furthermore, the source titles spanned 79 various sources. At the top was the International Journal of Production Topics with 13 publications, while 47 different sources titles contained only one publication each. This research offers an insight into fashion-industry literature, specifically on supply chain management. The statistics gathered from Web of Science might enlighten researchers, who want to publish a document in this area, about their publication options.

B. Cetinguc · E. Calik
Faculty of Engineering, Department of Industrial Engineering, University of Yalova, Yalova, Turkey
e-mail: basak.cetinguc@yalova.edu.tr

E. Calik
e-mail: ecalik@yalova.edu.tr

F. Calisir (✉)
Faculty of Management, Department of Industrial Engineering, Istanbul Technical University, Istanbul, Turkey
e-mail: calisirfet@itu.edu.tr

R. Rinaldi and R. Bandinelli (eds.), *Business Models and ICT Technologies for the Fashion Supply Chain*, Lecture Notes in Electrical Engineering 413,
DOI 10.1007/978-3-319-48511-9_2

Keywords Fashion industry · Supply chain management · Scientometric · Publication trends

1 Introduction

The need for success in supply chain practices in the fashion industry has attracted researchers and managers. The flexibility and responsiveness of firms determine commercial success or failure in fashion markets that require competitors to adapt to rapidly changing demand (Christopher et al. 2004). Moreover, the fashion system, as one of the global industries in a highly competitive world, has complex supply networks in terms of fragmentation and geographical dispersion of production activities and stakeholders (Macchion et al. 2015). In other words, it is often the case that clothing is designed in one country, manufactured in another, and sold in a third (Čiarnienėa and Vienažindienė 2014). Therefore, firms have redefined their business models in order to transform their relationships with supply-network members and to manage these chains attentively (Macchion et al. 2015). To orchestrate the integration and management of supply-chain members in the fashion industry becomes more complicated than in other sectors that have stable and predictable demand, such as the steel or chemical industry. The uncertain, complex, and dynamic nature of the fashion industry, characterized by short product life- cycles and great product variety, attracts researchers who work in the area of operations and supply chain management (Brun and Castelli 2008; Macchion et al. 2015). The goals of supply chain management in this industry are meeting the customer needs with increased variety, at the right time and in the right place (Azevedo et al. 2014). Members of fashion supply chains are mainly raw material suppliers, textile manufacturers, clothing manufacturers, retailers, and consumers (Azevedo et al. 2014).

Globalization and digitalization, as well as the growing demands for quick response and fast fashion, have been putting greater pressure on the fashion industry, so that supply chain management has become more strategically important than ever before (Iannone et al. 2015). The competition among retailers in the fashion industry is aggressive due to the recent success of fast fashion retailers (Čiarnienėa and Vienažindienė 2014). Supply-chain decisions have been affected by the dramatic shift in the scale and power of major retail buyers in the fashion market (Brun and Castelli 2008). Additionally, responding quickly to changing consumer needs and retail behaviors have played a crucial role in this industry (Macchion et al. 2015). Decreasing delivery times in the fashion supply chain is the main target of quick response (Čiarnienėa and Vienažindienė 2014). At the same time, the current trend of sustainable development in the economy, environment, and society has increased the awareness of a sustainability approach and sense of their importance in the fashion industry (Li et al. 2014). Thus, sustainable supply chain management in the fast fashion industry has become one of the prominent concepts in this area. Moreover, globalization of both sourcing and distribution has motivated the development of sustainable practices in fast-fashion supply chains (Turker and Altuntas 2014).

This scientometric study might shed light on the tendencies of supply chain management in fashion-industry literature since, different aspects and various kinds of concepts are being searched under the keywords of this literature. In the second section, we discuss how the data were gathered and which methodology was followed. The third section covers the findings, and, finally, in the last section, results are reported and discussed.

2 Data and Methodology

The data were collected from the Web of Science database. The databases that the Web of Science covered at the time were the Web of Science™ Core Collection, Derwent Innovation Index, KCI-Korean Journal Database, and SciELO Citation Index. "Fashion Industry" and "Supply Chain Management" were used as keywords to perform this research. Without time limitations the time bounds, it was found that a total of 123 publications related to supply chain management in the fashion industry were published up to January 2016. Conducting a scientometric analysis is similar to capturing the literature data up to that moment. So, this research covers the literature pervious to January 2016; any alteration or addition after that month was not taken into consideration by this research paper. Filters supported by the Web of Science were used to perform classification of the publications. Publication year, language, country, research area, documentation type, and source title are the main criteria of the classification.

3 Results

3.1 Publication Years

2000 was the first year of publication considered for supply chain content in the fashion industry, while sorting publications according to their publication years. Figure 1 shows the chronological order of the publications, and the most productive year was 2014 with 23 publications. By the year 2006, the trend in publications started increasing and never decreased below five publications after that year. One of the most productive years was 2008 by 12 publications. After 2011, the number of publications never dropped below ten. Since this research was conducted at the beginning of 2016, no publications for the year 2016 are included. Approximately 34% of the publications were published in the last 2 years.

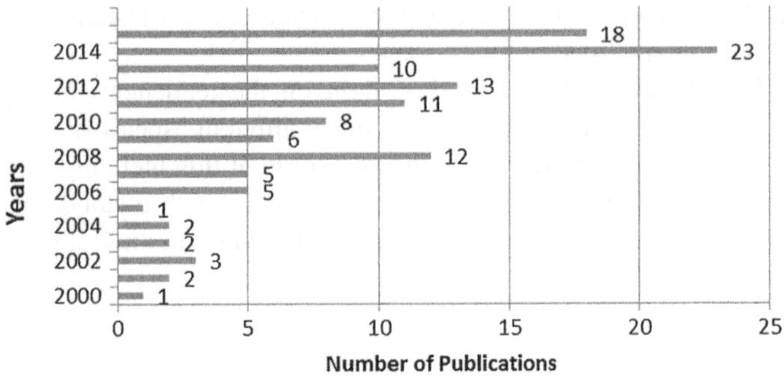

Fig. 1 Chronological distribution of publications

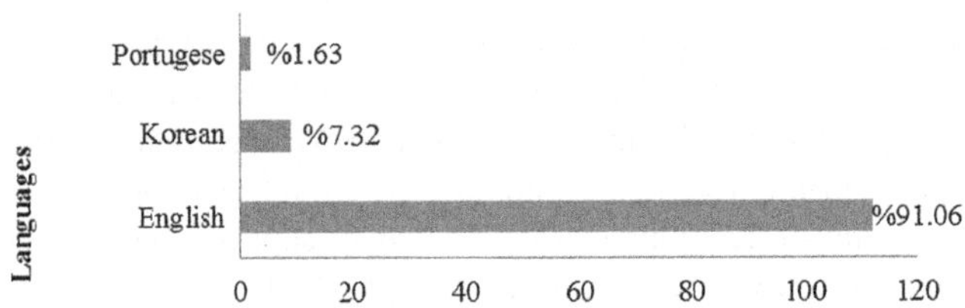

	English	Korean	Portugese
Number of Publications	112	9	2

Fig. 2 Number of publications by language

3.2 Languages

Three different languages were used in the publications related to supply chain management in the fashion industry. Figure 2 shows that the number of English-language publications is far more than those in Korean and Portuguese. 112 of 123 publications are written in English, whereas nine are in Korean and only two are in Portuguese. In order to understand whether this distribution is specific for supply chain management in the fashion industry, Web of Science database was used as a tool to identify the language distribution in supply chain management and in the fashion industry separately. It is found that publications were written in 17 different languages in the area of supply chain management area 18 different languages in the fashion industry in general. Nevertheless, English is the most popular language in both areas.

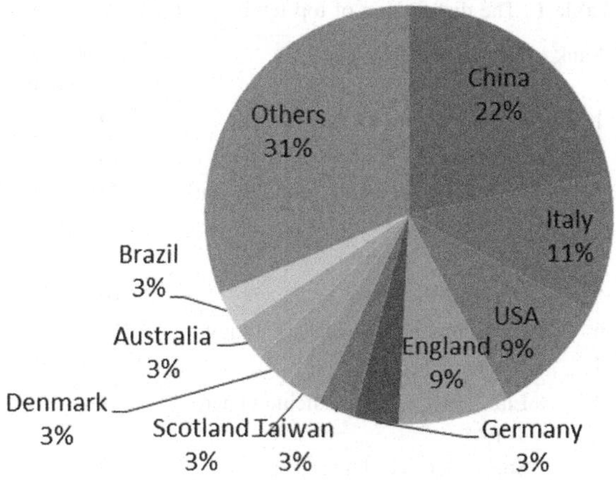

Fig. 3 Distributions of publications by country

3.3 *Countries*

In total, 33 different countries were the origin of the publications produced until January 2016. The distribution of countries can be seen in Fig. 3. China is the most productive country with 30 publications. In descending order, China, Italy, the USA, and England are the most productive countries, and 51% of the publications have been written in these countries. Five publications came from Germany, and four publications per country were generated by Australia, Brazil, Denmark, Taiwan, and Scotland.

Others, including Turkey, Spain, South Korea, Singapore, Norway, France, Finland, and Canada had three publications each; Romania, Netherlands, Lithuania, India had two publications per country; and Wales, Sweden, Sri Lanka, Saudi Arabia, Portugal, New Zealand, Japan, Ireland, Iceland, Greece, and Colombia had one publication. The sum of the classification by country totals 138 publications. It can be concluded that publishing companies of some publications probably collaborate with researchers from various different countries. Some other conclusions can be drawn from evaluating the distributions of languages and countries, for example, although China is the most productive country, Chinese does not appear in the list. Chinese researchers publish their publications in other languages. Similarly, Italy is the second most productive county, but no Italian-language publication has been published. Moreover, South Korea was the company location of only three publications in the databases, while a total of nine publications have originated from Korea.

3.4 *Research Areas*

Publications related to supply chain management in the fashion industry appeared in 17 different research areas. Since supply chain management is an interdisciplinary subject, there is a great diversity of research areas.

Table 1 The distribution of top ten research areas

Rank	Top 10 research area	Number of publications	% share in publications
1	Engineering	49	26.3
2	Business Economics	40	21.5
3	Operations Research Management Science	32	17.2
4	Computer Science	14	7.5
5	Materials Science	10	5.4
6	Environmental Sciences Ecology	10	5.4
7	Mathematics	7	3.8
8	Life Sciences Biomedicine Other Topics	7	3.8
9	Social Sciences Other Topics	5	2.7
10	Automation Control Systems	3	1.6

Table 1 shows the top ten research areas. Engineering is the first, by covering 26.3% of total publications. Business economics and operations research management science are the following top two research areas. After these three areas, there is a dramatically decrease in publications across research areas.

3.5 Document Types

Five types of documents are found related to supply chain management in fashion-industry literature. Articles are the most popular among the document types with 94 publications, followed by conferences with 18 publications. The remainder consists of reviews (8), patents (2) and editorial (1) publications. Figure 4 represents the percentage distribution of publications per document type.

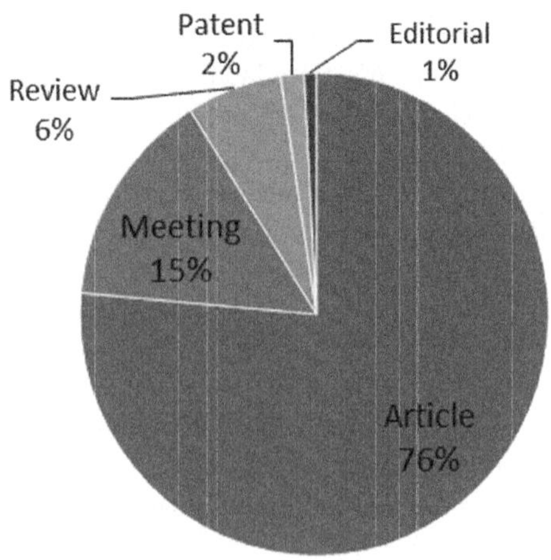

Fig. 4 Document type distributions of publications

Table 2 The distribution of top ten source titles

Rank	Top 10 source titles	Number of publications	% share in publication
1	International Journal of Production Economics	13	10.57
2	Production Planning Control	7	5.69
3	Mathematical Problems in Engineering	7	5.69
4	Sustainability	5	4.07
5	Journal of Fashion Business	4	3.25
6	The Research Journal of the Costume Culture	3	2.44
7	Supply Chain Management an International Journal	3	2.44
8	Journal of the Textile Institute	3	2.44
9	Journal of Business Ethics	3	2.44
10	International Journal of Operations Production Management	3	2.44

3.6 Source Title

There are 79 different source titles in the Web of Science Database related to supply chain management in the fashion industry. As can be seen in Table 2, the most popular journal is the International Journal of Production Economics, with 13 publications. Production Planning Control and Mathematical Problems in Engineering journals share the second place on the list with seven publications each. Fifty-seven source titles contain only one publication.

4 Discussion and Conclusion

This study is focused on the publications trend in supply chain management in the fashion industry. The Web of Science is used as a tool to collect the publications related to this topic until January 2016. The main objective of this study is to conduct a scientometric analysis of trends in supply chain management in the fashion industry. In order to perform the analysis, supply chain management and fashion industry were the keywords searched in the Web of Science search engine by topics, using the "AND" conjunction. Totally, 123 publications were gathered. Furthermore, publications were examined in terms of these criteria: publication years, countries, languages, research areas, document types, and source titles. The first publication related to the supply chain management in the fashion industry was published in 2000, and the year 2014 was the golden year of publications. It was found that three different languages were used in publications. In addition, the most productive country was found to be China with 30 publications, followed by Italy and the United States of America. Moreover, the most prominent research area was

Engineering with 26.3% and followed by Business Economics with 21.5% of all publications. Document types were divided into five different groups, and articles had the greatest share with 76% of publications. In terms of source title, International Journal of Production Economics is the most favorite journal in this area with 10.6%. Even though this study clarifies the publication trends in supply chain management in the fashion industry, there are few inevitable limitations. For instance, only the publications indexed in Web of Science database are included in this study. Furthermore, the ever-changing nature of publishing affects the number and distribution of publications.

In conclusion, some points revealed by this study are notable. Firstly, there is not any one journal that gathers under its roof of the supply chain management publications in the fashion industry. Researchers working on this subject could collaborate and publish a journal under the name of supply chain management practices in the fashion industry. Additionally, while the fashion industry also operates within Europe, there are not a sufficient number of publications originating from the many European countries. In this industry, sustainable fashion supply chain and fast-fashion supply chain management are new trends; hence these areas could be fruitful and exciting areas for researchers. For further research, databases apart from Web of Science could be used, and the results could be compared. This research was also done in January 2016. Henceforth, analysis could be conducted again, by following the same steps, in order to compare the results. Finally, the survey of research titles could be narrowed down more specifically, for example, to conduct scientometric analysis specifically in sustainable supply chain management in the fast-fashion industry.

References

Azevedo SG, Prata P, Fazendeiro P (2014) The role of radio frequency identification (RFID) technologies in improving process management and product tracking in the textiles and fashion supply chain. Woodhead Publishing, Cambridge

Brun A, Castelli C (2008) Supply chain strategy in the fashion industry: developing a portfolio model depending on product, retail channel and brand. Int J Prod Econ 116(2):169–181

Christopher M, Lowson R, Peck H (2004) Creating agile supply chains in the fashion industry. Int J Retail Distrib Manag 32(8):367–376. Available from Science Direct (12 January 2016)

Čiarnienė R, Vienažindienė M (2014) Agility and responsiveness managing fashion supply chain. Proc-Social Behav Sci 150:1012–1019

Iannone R, Martino G, Miranda S, Riemma S (2015) Modeling fashion retail supply chain through causal loop diagram. IFAC-PapersOnLine 48(3):1290–1295

Li Y, Zhao X, Shi D, Li X (2014) Governance of sustainable supply chains in the fast fashion industry. Eur Manag J 32(5):823–836

Macchion L, Moretto A, Caniato F, Caridi M, Danese P, Vinelli A (2015) Production and supply network strategies within the fashion industry. Int J Prod Econ 163:173–188

Turker D, Altuntas C (2014) Sustainable supply chain management in the fast fashion industry: an analysis of corporate reports. Eur Manag J 32(5):837–849

Cross-Cultural Research for Luxury Fashion Brands in the Chinese Market: A Review of Long- Versus Short-Term Orientation in National Culture Dimensions

Ruichen Lu, Yi Wang and Richard Kennon

Abstract This study reviews the culture dimensions and mainly summarizes and provides recommendations according to the fifth dimension of Hofstede cultural model. The target of this paper is to support the cross-cultural research for luxury fashion brands in the Chinese market via a review of long- versus short-Term orientation in national culture dimensions. The research method of this study is documentary analysis. The three future directions for luxury fashion brands in the Chinese market are presented including customization, improving the VIP experience, and innovative media technology.

Keywords Cross-cultural research · National culture · Dimensions of national culture · Luxury fashion · Chinese luxury market

1 Introduction

Cross-cultural research is widely used in multinational brands' operation and management. The growth of demand for luxury fashion brands in Asia's emerging economies with higher levels GDP than before (Halepete 2011), especially in China, has driven the broadening of the luxury market (Li et al. 2012). The Chinese, luxury

R. Lu (✉)
School of Materials, The University of Manchester, Sackville Street Building-B36A, Manchester M13 9PL, UK
e-mail: ruichen.lu@postgrad.manchester.ac.uk

Y. Wang
School of Materials, The University of Manchester, Sackville Street Building-C51G, Manchester M13 9PL, UK
e-mail: yi.wang-2@manchester.ac.uk

R. Kennon
School of Materials, The University of Manchester, Manchester M13 9PL, UK
e-mail: richard.kennon@manchester.ac.uk

R. Rinaldi and R. Bandinelli (eds.), *Business Models and ICT Technologies for the Fashion Supply Chain*, Lecture Notes in Electrical Engineering 413,
DOI 10.1007/978-3-319-48511-9_3

fashion market should be explored to promote global collaboration between transnational companies, as well as to contribute to their internationalization (Laskowska-Rutkowska 2009). Changes, such as consumer preference, in the market are influenced by many factors like culture diversity (Halepete 2011), population, educational status, technological level, etc. (Rauch and Meier 2000). This paper reviews the culture dimensions and mainly summarizes and provides recommendations according to the fifth dimension of Hofstede cultural model. This paper is divided into three parts. The first part is the review of culture, including concepts of culture and national culture. The second part is an overview of Hofstede's cultural model and a review of fifth dimension. The final part is an analysis of the current Chinese market for luxury fashion brands and its likely future direction.

2 The Concept of Culture

Culture is defined as the personality of a society (Solomon et al. 2014, p. 549) or a "shared meaning system" (Fischer 2009). Culture is not a personal attribute but refers to a mass of people with similarities of education and life history (Hofstede et al. 2010). Aspects of culture include values, customs, symbols, and language (Solomon et al. 2014).

Hofstede and his associates affirmed a positive relationship between culture and economic growth, based on their published statistical evidence (Hofstede 1980). While Yeh and Lawrence (1995) averred that culture is not a sufficient reason to support the growth of the economy, they believed that other factors determine economic growth.

3 National Culture

Frequently, national culture is considered in international marketing, when global corporation managers hope to enter new overseas markets (De Mooij 2015). Steenkamp et al. (1999) suggested that an individual's innovativeness is affected by national culture. This paper uses the technique of hierarchical linear modeling to distinguish between degrees of personal values and national culture. In short, the environment of the national culture is able to impact on consumer preference. Understanding changes in national culture helps multinational corporations to explore the consumption structure in target markets (Craig and Douglas 2005).

Only cross-cultural research, using data on the national level, is considered, because of the paucity of data on differences within nations (Hofstede and Minkov 2011). Although there are heterogeneous events across some groups of nations, the diversities between nations tend to be much larger than within any one (De Mooij 2014).

4 Hofstede's Five-Dimensional Model of National Culture

Cultural models are rooted in cross-cultural research and then developed in cross-cultural management. One of the most famous cultural models is Hofstede's cultural model, established in 1973 by Geert Hofstede, a Dutch scholar (De Mooij 2015). This model includes five dimensions–power distance, individualism versus collectivism, masculinity versus femininity, uncertainty avoidance, and long- versus short-term orientation (Hofstede 1991).

5 Long- Versus Short-Term Orientation

Greet Hofstede's first in-depth examination assumed a model of national cultures with the first four dimensions (Hofstede 1980). In following research, the fifth dimension was added to this model by Hofstede and his associates (Hofstede 1991). The fifth dimension is rooted in research into students' values based on the Chinese Value Survey (CVS), which was initiated by Michael Harris Bond. Participants came from 23 countries around the world (Minkov and Hofstede 2012). The fifth dimension has two poles, one positive (long-term orientation) and one negative (short-term orientation), each with four variations (Table 1). Since all eight items can be found in Confucius' teaching, the fifth dimension was first named "Confucian Work Dynamism" by Michael Bond (Chinese Culture Connection 1987). Then Hofstede (1991) renamed it "Long-Term versus Short-Term Orientation" (LTO) when he accepted this dimension.

Since the emergence of this dimension, it has been questioned, critiqued, and even ignored by many scholars. Franke et al. (1991) indicated that a certain correlation exists between the fifth dimension and an individualism-versus-collectivism dimension. Strong interrelationship between them is also found by Yeh and Lawrence (1995). In short, it is possible that Confucian Dynamism cannot be a dimension alone. The fifth dimension is even omitted by Triandis in his review of Hofstede's culture dimensions (Hofstede 1993).

Table 1 Long-term versus short-term orientation (Hofstede 1991; Fang 2003; Minkov and Hofstede 2012)

On the positive pole:
Persistence (perseverance)
Ordering relationships by status and observing this order
Thrift
Having a sense of shame
On the negative pole:
Personal steadiness and stability
Protecting your "face"
Respect for tradition
Reciprocation of greetings, favors, and gifts

Table 2 The merits and demerits of long-term versus short-term orientation

Merits	Demerits
Achieving supply chain management effectively (Mentzer et al. 2001, p. 10)	Correlation between fifth dimension and individualism-versus- collectivism dimension (Franke et al. 1991)
Long-term competitive power in supply chain (Chen and Paulraj 2004, pp. 138–139)	Confusion from the fifth dimension's name (Hofstede 2001)
The foundation of trust (Cannon et al. 2010)	Philosophical drawbacks of the fifth dimension (Fang 2003)

Moreover, the term fifth dimension is revised, from Confucian Dynamism to long-term versus short-term orientation. Since the initial name, sounding exclusively Chinese, evokes a sense of strangeness by many readers in Western countries, the terms "Confucian Dynamism" and "Long-Term versus Short-Term Orientation" are not used any more interchangeably by Hofstede (2001). The reason for dropping its use is recent evidence of Eastern European countries' high scores in Long-Term Orientation Index (Minkov and Hofstede 2012). The absoluteness inherent in the terms positive and negative, two aspects of the fifth dimension, is also a philosophical drawback. What's more, the four aspects on the negative pole do not have negative meanings in Chinese phrases such as "wenzhong"–personal steadiness and stability (Fang 2003).

Although the fifth dimension is questioned continually, it still has broad application and longevity, especially in supply chain management. First of all, long-term relationships can make supply chain management effective and then promote cooperation between partners (Mentzer et al. 2001; Santos et al. 2012). Moreover, in the whole supply chain, a long-term relationship implies a kind of long-term competitive power (Chen and Paulraj 2004). This could decrease the number of suppliers and then reduce the complexity in a long-term cooperation (Stevens 1989). Finally, long-term activity is the foundation of trust between buyers and suppliers (Cannon et al. 2010). Thereby, high-level information sharing could be achieved in order to reduce cultural differences and resistance to collaboration (Cheung et al. 2010).

The following diagram displays the merits and demerits of Long-Term versus Short-Term Orientation (Table 2).

6 Current Situation of the Chinese Luxury Market

As an emerging market, China is one of the BRICs countries with fast economic growth and a high-level of GDP (Halepete 2011). China is focused on the international fashion industry, especially luxury fashion brands. From ancient times until today, luxury is not an unfamiliar concept to consumers in Chinese markets with traditional, social structure. Due to the influence of Confucian culture, one is expected to undertake obligations in both the professional and family contexts (Atwal and Bryson 2014).

6.1 *Superiorities of the Chinese Market*

First of all, gift-giving plays a crucial role to show respect not only for one's superiors and leaders but also gratitude to parents and elders (Atwal and Bryson 2014). Luxury has been rooted in Chinese social customs and habits to maintain complicated relationships, even today, through the form of gifts to counter-gifts (Tsai 2008). Second, the size of its cities determines the global influence of China. More than 200 cities with 100 million or more people will have emerged in China by 2030 (Atwal and Bryson 2014). At the same time, the main luxury consumer group, the middle class, is growing fast (Wang 2010). This condition means that Chinese markets will be further broaden, rather than be concentrated in several metropolises, such as Shanghai and Beijing.

Moreover, Confucian heritage emphasizes the importance of family. Most Chinese families not only respect traditional culture but also apply it in modern society. The children have become the center of the family due to the one-child policy. Thus luxury for children becomes a new trend (Atwal and Bryson 2014). This constitutes a big difference between the Chinese and other countries' markets. These children receive a good education and grow up in wealthy family environments. They have their own understanding of luxury fashion brands and buy their favorites spontaneously as self-rewards. Therefore, these characteristics lead to a younger-age trend of luxury consumers in China (Halepete 2011), where close to half of luxury-fashion consumers are less than 35 years old (Atsmon et al. 2011). This condition suggests that there is huge potential in the future Chinese market.

6.2 *Challenges of the Chinese Market*

On the other hand, future challenges of the Chinese market should not be ignored. High taxes on luxury fashion products are a big problem in China (Chang 2013), because they result in higher prices for luxury fashion products in the Chinese market. In order to avoid paying higher prices, many consumers choose to buy products in foreign destinations, instead of buying in their domestic market (Atsmon et al. 2012). This phenomenon results in an unpredictable and complicated Chinese market. The physical stores of western luxury fashion brands in China mainly play a role in increasing awareness of these brands among Chinese citizens.

Due to the huge Chinese potential, luxury fashion brands in China expand and rise rapidly. But competitive activities limit the profit margins of most brands. In recent years, the numbers of potential Chinese consumers have been rising steadily

Table 3 The pros and cons in the Chinese luxury market

Pros	Cons
Gift-giving culture (Atwal and Bryson 2014)	Higher price of luxury products in Chinese market (Chang 2013)
City size and middle class increase (Wang 2010)	Too many competitors in Chinese market (KPMG 2013)
The younger-age trend of luxury consumers (Halepete 2011)	Faltering demand for flashy goods (Atwal and Bryson 2014)
	Anti-corruption actions against Chinese government officials (Atwal and Bryson 2014)

and the market has become crowded (KPMG 2013). Furthermore, the demand for flashy goods that carry a prominent brand logo is faltering because of aesthetic fatigue and market maturity. In addition, the recently installed Chinese government has started to prosecute corrupt officials and attempted to control their lavish life style, beginning in 2012. Product sales of many luxury fashion brands, such as Cartier and Channel, have declined dramatically (Atwal and Bryson 2014).

Therefore, having a good understanding of Chinese culture and conditions is the key success factor to international companies when they start their investments in China. The following diagram summarizes the pros and cons of the Chinese luxury market (Table 3).

6.3 Future Direction

Looking ahead, customization will be the main trend of luxury fashion brands in the Chinese market, because more and more younger luxury consumers pursue experiential luxury. At the same time, they hope to re-define luxury rather than just follow it (Bain and Company 2015). Improving the VIP experience is conducive to long-term orientation between consumers and retailers and this then achieves exclusivity, for example, Burberry's Bespoke service directed at trench coats (Ahrendts 2013). In addition, innovative media technology is an important tool to extend e-commerce platforms. Luxury fashion companies can use Chinese popular social media, such as Wechat and Weibo, to strengthen their influence and popularity in the Chinese market (Fung Business Intelligence Centre 2013).

7 Conclusion and Recommendations

During recent years, there is a widespread use of cross-cultural research in international markets due to the speeding up of global integration. How to explore cultural differences between domestic markets and international markets is a huge challenge confronting multinational companies. Firstly, ethnocentrism should be avoided by researchers during their investigation. Secondly, a good understanding of the cultural concept and a highly applicable cultural model are able to enhance the success of any cross-cultural research (Yaprak 2008; De Mooij 2015). Certainly with the rise of emerging markets, some cultural models should be improved or reconstituted. Finally, the complexity of international markets, as in the Chinese market, cannot be ignored. In other words, research into natural culture should not be isolated or detached and then simply defined as positive or negative.

References

Ahrendts A (2013) Burber's CEO on turning an aging British icon into a global luxury brand. Harvard Business Review, pp 39–43

Atsmon Y, Dixit V, Leibowitz G, Wu C (2011) Understanding China's growing love for luxury. http://solutions.mckinsey.com/insightschina/_SiteNote/WWW/GetFile.aspx?uri=:/insights-china/default/en-us/aboutus/news/Files/wp2055036759/McKinsey%20InsightsChina%20-%202011%20Luxury%20Consumer%20Report_7fbe9eff-0bfd-4cea-8bc5–2dff874db6d0.pdf

Atsmon Y, Ducarme D, Magni M, Wu C (2012) Luxury without borders: China's new class of shoppers take on the world. The McKinsey Chinese luxury consumer survey. http://www.mckinseychina.com/wp-content/uploads/2012/12/the-mckinsey-chinese-luxury-consumer-survey-2012-12.pdf

Atwal G, Bryson D (2014) Luxury brands in emerging markets, 1st edn. Palgrave Macmillan

Bain and Company (2015) China's luxury market shrinks in 2014 as luxury brands adapt to shifting consumer preferences. http://www.bain.com/about/press/press-releases/chinas-luxury-market-shrinks-in-2014-press-release.aspx

Cannon JP, Doney PM, Mullen MR, Petersen KJ (2010) Building long-term orientation in buyer–supplier relationships: the moderating role of culture. J Oper Manag 28(6):506–521

Chang GG (2013) Chinese shunning luxury stores at home, buying abroad, Forbes. http://www.forbes.com/sites/gordonchang/2013/05/05/chinese-shunning-luxury-stores-at-home-buying-abroad/

Chen IJ, Paulraj A (2004) Understanding supply chain management: critical research and a theoretical framework. Int J Prod Res 42(1):131–163

Cheung MS, Myers MB, Mentzer JT (2010) Does relationship learning lead to relationship value? A cross-national supply chain investigation. J Oper Manag 28(6):472–487

Chinese Culture Connection (1987) Chinese values and the search for culture-free dimensions of culture. J Cross Cult Psychol 18(2):143–164

Craig CS, Douglas SP (2005) International marketing research. Wiley, Chichester

De Mooij M (2014) Global marketing and advertising. understanding cultural paradoxes, 4th edn. Sage, Los Angeles, CA

De Mooij M (2015) Cross-cultural research in international marketing. Clearing up some of the confusion. Int Mark Rev 32(6)

Fang T (2003) A critique of Hofstede's fifth national culture dimension. Int J Cross Cultural Manag 3(3):347–368

Fischer R (2009) Where is culture in cross-cultural research? An outline of a multilevel research process for measuring culture as a shared meaning system. Int J Cross Cultural Manag 9(1): 25–49

Franke RH, Hofstede G, Bond MH (1991) Cultural roots of economic performance: a research note. Strateg Manag J 12:165–173

Fung Business Intelligence Centre (2013) Luxury market in China: huge growth potential ahead. http://www.funggroup.com/eng/knowledge/research/china_dis_issue109.pdf
Halepete J (2011) Retailing in emerging markets. Fairchild, New York
Hofstede G (1980) Culture's consequences: international differences in work-related values. Sage, Beverly Hills, CA
Hofstede G (1991) Cultures and organizations: software of the mind. McGraw-Hill, London
Hofstede GH (1993) [Book review] cultures and organizations, software of the mind. Adm Sci Q 38:132–134
Hofstede G (2001) Culture's consequences: comparing values, behaviors, institutions, and organizations across nations. Sage, Thousand Oaks, CA
Hofstede G, Hofstede GJ, Minkov M (2010) Cultures and organizations: software of the mind, 3rd edn. McGraw Hill, New York, NY
Hofstede G, Minkov M (2011), Is national culture a meaningful concept? Cultural values delineate homogeneous national clusters of in-country regions. Cross-Cultural Res XX(X):1–27
KPMG (2013) Global reach of China luxury. http://www.kpmg.com/CN/en/IssuesAndInsights/ArticlesPublications/Documents/Global-Reach-China-Luxury-201301.pdf
Laskowska-Rutkowska A (2009) The impact of national and organizational culture on the cooperation of firms: a supply chain perspective. J Intercultural Manag 1(2):5–16
Li G, Li G, Kambele Z (2012) Luxury fashion brand consumers in China: perceived value, fashion lifestyle, and willingness to pay. J Bus Res 65(10):1516–1522
Mentzer JT, Dewitt W, Kleeber JS, Min S, Nix NW, Smith CD, Zacharia ZD (2001) Defining supply chain management. J Bus Logist 22(2): 1–25
Minkov M, Hofstede G (2012) Hofstede's fifth dimension: new evidence from the World Values survey. J Cross-Cultural Psychol 0022022110388567
Rauch JE, Meier GM (2000) Leading issues in economic development. Oxford University Press, New York
Santos et al. (2012) Cultural dimensions of agile supply chain management. POMS 23rd annual conference
Solomon MR, Dahl DW, White K, Zaichkowsky JL, Polegato R (2014) Consumer behavior: buying, having, and being. Prentice Hall, Upper Saddle River, NJ
Steenkamp J-BEM, Ter Hofstede F, Wedel M (1999) A cross-national investigation into the individual and national cultural antecedents of consumer innovativeness. J Mark 63(2):55–69
Stevens GC (1989) Integrating the supply chain. Int J Distrib Mater Manag 19(8):3–8
Tsai J (2008) La Chine et le luxe. Odile Jacob, Paris
Wang H (2010) The Chinese dream: the rise of the world's largest middle class and what it means to you. Bestseller Press, New York
Yaprak A (2008) Culture study in international marketing: a critical review and suggestions for future research. Int Mark Rev 25(2):215–229
Yeh RS, Lawrence JJ (1995) Individualism and Confucian dynamism: a note on Hofstede's cultural root to economic growth. J Int Bus Stud 655–669

See Now Buy Now: A Revolution for Luxury Supply Chain Management

Alessandro Brun, Cecilia Castelli and Hakan Karaosman

Abstract Fashion collections used to be for the eyes of editors, buyers and journalists. Yet, digital technology and IT transformation have opened this once-enclosed world by transforming it into an open access arena. Fashion trends are immediately exposed through social media; however, in the current fashion system, there is a wait of at least six months between runway shows and retail availability. Some might wonder if today's consumers could still wait for what a traditional fashion calendar offers. To this end, the entire fashion system is about to be reconfigured. See Now Buy Now (SNBN) emerges as a business model to make a fashion revolution by putting the fashion items immediately on sale after runway shows. Nevertheless, it is extremely vague in terms of what this movement will constitute for those involved in supply chain (SC) operations. Thus, it is time to interrogate if this movement could build a new system with improved production processes to fulfil demand more efficiently. Looking at the SNBN revolution with the standpoint of academic researchers, this study aims to take a closer glance at what is bound to change at SC and operations level. Furthermore, it is set to identify what is required to enable SNBN in terms of SC structure and capabilities.

Keywords See now buy now · Luxury fashion industry · Supply chain management · Luxury fashion operations · Fashion revolution

A. Brun · H. Karaosman (✉)
Department of Management, Economics, and Industrial Engineering,
Politecnico di Milano (POLIMI), Milan, Italy
e-mail: hakan.karaosman@polimi.it

A. Brun
e-mail: alessandro.brun@polimi.it

C. Castelli
The European House—Ambrosetti, Milan, Italy
e-mail: Cecilia.castelli@ambrosetti.eu

R. Rinaldi and R. Bandinelli (eds.), *Business Models and ICT Technologies for the Fashion Supply Chain*, Lecture Notes in Electrical Engineering 413,
DOI 10.1007/978-3-319-48511-9_4

1 Introduction: The Current Fashion-Shows Scenario and the See Now Buy Now Revolution

The relevance of fashion weeks in its current format is being questioned and there appears to be a need for a road map for the future of luxury fashion. The Council of Fashion Designers of America (CFDA) is rethinking about the fashion week format (Bobila 2015). Potential developments including more intimate presentations for the industry and runway shows tailored to retailers and consumers are sought. In this vein, the CFDA has launched an online-only digital resource for the fashion week calendar and is already re-evaluating the traditional twice-yearly format of New York Fashion Week during which designer collections are displayed for industry insiders six months in advance. Chairman of the CFDA and designer Diane Von Furstenberg addresses that designers, retailers, and everybody complains about the shows at the fashion week in its current format (Lockwood 2015). Therefore, a need emerges to create fashion shows geared more towards 'the consumer experience'. For instance, Riccardo Tisci brought Givenchy to New York Fashion Week for the first time, in an alignment with the opening of its massive flagship on Madison Avenue in Manhattan in February. Furthermore, Tisci opened the show to the general public, gave away 820 tickets, including 280 tickets for local fashion students and 100 for locals. Similarly, fall/winter 2016/2017 fashion weeks in New York, London, Milan and Paris recently witnessed a change and some designers already change their approaches to display fashion collections.

Given that social media has a pivotal role in marketing, the principal goal now aims to take the analogue business formula and translate it into a new format for today's digital consumers. In the age of social media, Burberry and Tom Ford announced a revolutionary SNBN movement. The SNBN aims to put the fashion items on sale immediately subsequent to runway shows. Both brands are reportedly launching the SNBN in their fashion shows scheduled for September. Burberry CEO Christopher Bailey states that 'Instagramming, live-streaming and showing the collections online means that brands can't expect a consumer to tie-in with a traditional fashion calendar'. Relatedly, 'In a world that has become increasingly immediate, the current way of showing a collection four months before it is available to consumers is an outdated idea and one that no longer makes sense' Tom Ford states in a release (Mower 2016). In the current system, consumers need to wait at least six months; nevertheless, it is visible that the fashion system is about to be reconfigured. Along with Burberry and Tom Ford, smaller brands, such as Rebecca Minkoff and Vetements, aim to forsake the traditional fashion calendar and to show collections that will be immediately available for purchase (Wischhover 2016). Some other brands, including Prada, Diane von Furstenberg and Monique Lhuillier have made similar moves, but rather on a smaller scale. For example, Monique Lhuillier addresses that when consumers see an image, they want to have it immediately; therefore, the designer created five looks for her fall/winter 2016 collection that were ready for an instant purchase.

Correspondingly, Thakoon announced its acquisition by Bright Fame Fashion whose strategy is the SNBN. Further, Chris Benz has accomplished a radical brand makeover through SNBN phenomena. Benz, who took the creative leadership of American sportswear label Bill Blass in 2014, has launched a new website with a live stream of the house's archives, and collections will not be revealed till they are available for sale and new products will be rolled out continuously on a monthly basis (Mau 2016). LVMH prize winner Thomas Tait will also opt to schedule private appointments with buyers and press (Fernandez 2016). Tommy Hilfiger, relatedly, has introduced a virtual experience in its stores. The brand has started showing collections to buyers and the press in advance and keeping the runway shows for the consumers when the clothes actually hit the store. Given that the cultural conversation is provocative, and it is incorporating change, new ideas, innovation and new channels, the SNBN could be considered a total embrace of innovation. To this end, the chief executive of the British Fashion Council, Caroline Rush, told the New York Times that this strategic move displays an excellent leadership in driving the fashion agenda forward. Burberry, having its own factory for trench coats, has about 70% of its revenue coming from its own retail stores and the purchasing power to create stock allows promptness for a simultaneous showing and selling. In February, Burberry showcased contrasting metallic dresses, snake-skin trench coats and embellished shirts in rich colour palettes all of which were already in-store for people to see, touch, and feel. Further, delivery was set to take place as quickly as possible upon the orders.

Nonetheless, although the fashion industry needs to evolve, one rule does not seem to fit everybody. Not everyone in the fashion business could gear up for such a configuration. There is a mounting concern among smaller designers, depending on wholesale partners for distribution and retail, addressing that they would be unable to compete. Some brands already expressed that more time is needed to prepare the collections for wholesale and they might end up with extra inventory unless they were able to measure customers' reactions. Jackie Lee, a Korean designer based in London, stresses that as a smaller designer brand, they need time to get orders from buyers, put them into production, wait for the fabric and deliver the final product (Hoang 2016). Designers often rely on advance orders from retailers to align their production costs and decide how many pieces to order. Nonetheless, if they need to produce an entire collection to be sold right after the show, the process might become uncontrollable since 'an intense guesswork' is required. This also means that funds must be enough to get the production done immediately. An eco-conscious designer, Laura Siegel, who currently produces clothes only when she gets orders from retailers, says that if this become the norm it couldn't work (Maddeaux 2016). Small brands do not own their production facilities and they often share factories with other brands. Thus, even if a small designer would manage to finance his production costs and predict what will be sold, they still would not turn around collections as quick as large global retailers. It is worth stressing that products might not get ready because factories also work with other designers, and there might appear to be a problem for factories to prioritize their clients. Furthermore, if a brand uses artisans around the world to produce the garments, or when products are hand-sewn, the process is unlikely to speeded up to reduce leadtime. In this sense, Carlo Capasa, the head of Italy's national chamber of fashion, addresses that time is needed in order to make a creative collection. From the SC point of view, on the other hand, it is still unclear where SNBN will lead to for those involved in operations. Additionally, French designers do not seem as embracing as their British and American contenders. The administrative body of French fashion designers opted to reject the schedule change (Gonzales 2016). It was further stated that the wait between collections showcased on the runways and retail availability create excitement, whereas the SNBN claims that consumers lose interest in this long wait. A discussion is lead toward brand positioning. In fact, Ralph Toledano, the president of the federation, further addresses that the new schedule might work for marketing driven brands; however, Paris is not set to compromise the quality of work by French fashion designers. Chanel's creative director Karl Lagerfeld also reports that the iconic brand has already put items on sale from the pre-collection.

Despite the opposition, the SNBN is seen a way to benefit emerging designers if they could find ways to have closer relationships with consumers and figure out what is wanted. Besides, this new luxury paradigm is likely to challenge fast fashion. Fast fashion retailers, known to repeat runway trends without competition, will not be able to profit of the traditional six-month gap between catwalk shows and retail availability The SNBN movement is becoming a promising topic for a further discussion in SC management (SCM). Hence, it is time to interrogate if this movement could build a new system with improved production processes to fulfil demand more efficiently in luxury fashion SCM.

2 Research Objectives

Looking at the SNBN revolution with the standpoint of academic researchers, we were amazed by the number of research questions still unaddressed (let alone answered to), which could keep the academic community busy and bustling in the next few years. This being a very preliminary paper on the topic, we will focus on just two main questions:

- RQ1: the SNBN is a revolution for the market and for the fashion shows business; what is going to change at the SC and Operations level? What will the impact on the SC practices be?
- RQ2: what are the prerequisites in terms of SC structure and capabilities to enable SNBN? Which could be the hindrances—e.g. are there specific SC configurations making the SNBN more difficult or altogether impossible?

3 Research Methodology

To further establish this new territory, the objective of this research is twofold. Firstly, it offers results acquired from a quantitative study conducted on the distribution challenges of luxury fashion SCs*. Secondly, it provides an original contribution coming from four case studies carried out to conceptualize a framework for the SNBN.

3.1 Sampling

Premium quality and excellence should be pursued along luxury fashion SCs. However, today's global economic uncertainty, growing business complexity, fast fashion, the leadtime between fashion shows and retail availability, and the pivotal role of social media gradually challenge the luxury fashion. The luxury fashion emerges as a quite informative industry to further investigate the SNBN. With the unit of analysis being a fashion company, this research implements the principles of theory building based on case studies (Yin 1994). Given that we explore a relatively new research area, that is the SNBN, case study research has been deemed appropriate. Relatedly, a theoretical sampling approach was taken into account. Comparability of the cases was ensured through the introduction of a number of selection criteria. Contingency effects such as cultural differences on strategy deployment were reduced by a focus on a single country. Italy was chosen due to its distinctive significance for the luxury industry in terms of textile production and apparel manufacturing. E.g. a total number of companies producing and supplying leather account for 17662 worldwide, yet up to 43% of the global leather supply comes from Italy (retrieved from http://www.europages.co.uk). Italy thus emerges as one of the most significant countries for luxury fashion SCs. To fully investigate the SNBN impact on a SC through buying firm-supplier-subcontractor relationship, two SC levels were focused, namely apparel manufacturer (1st level) and textile producer (2nd level). Four suppliers located in Lombardy region of Italy were

Table 1 Company characteristics

	Products	Level in SC	Employees	Production process
Case A	Cotton products	1st level	45	Knitting, cutting, pressing, quality control
Case B	Cashmere products	1st level	41	Knitting, cutting, pressing, quality control
Case C	Silk products	1st level	50	Modelling, cutting, assembly, quality control
Case D	Leather goods	2nd level	180	Design, dyeing, finishing, quality control

approached to capture a range of characteristics. Eventually, data collected from four companies were utilized. Table 1 profiles the participants in this research.

3.2 Data Collection and Analysis

A structured analysis was conducted to ensure a high degree of consistency in data collected from each company. Data were collected using interviews, archival materials and documents. In 2016, the research team held the interviews for which a semi-structured interview protocol was developed. The protocol called for interviews with the top managers in charge of production. It must be addressed that, job titles vary in small and medium sized companies. While the research design called for expertise, respondents had multiple responsibilities. This reduced the number of interviews, but increased the depth of the conversation. Each interview continued from 60 and 90 min. Archival documents, on the other hand, were reviewed to triangulate the responses obtained during the interviews. Data collection stopped when saturation was achieved. Coding, which was based on the transcripts, interviewer notes and secondary data, was completed via a multistep iterative process. Firstly, within-case analysis was conducted by which the potential impact of the SNBN on operational and business performance was gained. Following within-cases, cross-case analysis was employed to identify common themes. Eventually, a coherent understanding of the development of the SNBN and the potential consequences emerged. The findings are presented in the next section.

4 Results

4.1 Distribution Challenges in Luxury Fashion SCs

The Politecnico di Milano research team carried out a survey on "Distribution Challenges" in 2015 addressing the current practices in luxury SCM. That research was not meant to analyse the impacts and implications of the SNBN; yet, we analysed the research results in hindsight, results trying to figure out some links with the SNBN revolution. Some interesting insights emerged, which shed some light on the most critical processes. Results address that the main distribution flows are usually measured over a period of time, e.g. a season or a year. Luxury fashion companies measure and quantify (i) the input flow, (ii) value of the goods entering the warehouses, (iii) value of the produced but not delivered goods, (iv) value of the goods distributed to the retail network, (v) value of the goods distributed to the wholesale network, (vi) value of the goods distributed to online, sales and re-sales output flows. However, the distribution phase becomes frantic. Due to the system complexity and know-how required, the majority of our sample was found outsourcing their distribution activities. The survey results reveal that almost 80% of the activities associated with online channel logistics are outsourced. Furthermore, luxury fashion companies entirely outsource their value added activities (100%), while transport is similarly fully outsourced. Warehousing activities are outsourced up to 50%; yet, interestingly, online channel information systems show an insourcing trend. It emerges that outsourcing might not be per se a competitive advantage but it could be a cost/effectiveness tool as long as the company possesses the control over the entire logistic process.

Customer demand is an erratic parameter in luxury fashion SCs since the number of product configurations is very high while the demand is not certainly known. Thus, quantifying the expected target service level becomes an overarching importance to design the distribution chain. Nevertheless, not all luxury fashion companies are equipped to define a clear target. Recently, overproduction is being held within the industry. To better manage customer demand and reduce overproduction; forecast optimization, free stock management, fast replenishment and overproduction optimization emerge as potential instruments to deploy. The analyses further show that new customer experiences appear in the luxury markets. Even though the luxury fashion companies are not fully prepared to fulfil the new customer demands, results reveal that luxury customer growingly seeks real time replenishment, virtual stock and CRM best practices. In the vein of the SNBN, channel integration (online, retail), markets integration and timely services (same day delivery) appear as key areas of implementation for luxury fashion SC processes.

Findings reveal that the utilization of free stock is still at its infant phase, as the percentage of free stock ranges between 0 and 30% of all sales, meaning that no free stock is a common situation. A more accurate approach would require separating order collection from order allocation, and gross demand from net demand. The survey results, on the other hand, address that the most common replenishment frequency is on a weekly basis. To enable a fast replenishment strategy with high frequency and high flexibility, a robust real sales data collecting process should be in place with adequate IT systems. It was revealed that the initial orders allocation (push system) and the limited use of free stock are reflected into the replenishment order model, resulting in the stores having the biggest responsibility. Only few companies use automatic replenishment, which requires visibility of the stock in a single store. It emerges that introduction of a multi-echelon distribution system could increase flexibility and efficiency. Further, it could also allow the application of an outsourcing strategy, consistent with the competitive advantage. When a multi-echelon distribution is in place, the second important step to tackle the uncertainty of sales forecasts could be the adoption of a free stock. Its impact on performances could achieve higher levels when integrated with a high frequency daily replenishment process (fast replenishment). Companies aiming to get a competitive advantage through distribution must optimize overproduction level through system modelling, and to this end using a non-competitive benchmarking with other sectors could assist.

4.2 The SNBN in Italy: Where Do Italian Luxury Fashion Companies Stand?

Having a flexible SC network is required to implement the SNBN. In the current system, only 10 items are released for luxury fashion shows; however, if the SNBN becomes a transformative movement, up to 1000 items would have to be realized and prepared to allow the buying firm for an immediate purchase, the owner of Case C revealed. Therefore, production and delivery flexibility are required to adapt the change. Case B, on the other hand, finds the SNBN rather challenging. It was addressed that they receive the orders right after fashion shows and leadtime is about two months to deliver the final products to the client. However, if the SNBN is implemented forecasting must work extremely accurate to prevent any potential overstock problems. Case D, located at the second level of the chain, exposed that the SNBN is not difficult to implement if a company has a flexible manufacturing process. In this vein, it was addressed that the company have stocks of 100 metres of textile for each colour and for each product type, which allow the company to meet high and rapidly changing customer demand. However, if the client requires a special textile featuring specific colours, patterns and/or texture, an average of 10 days are needed to finish the product. As leather requires some time to get dried and some chemicals are utilized in the production process, reducing the production time seems challenging. Case A, instead, appears to be a peculiar case for the SNBN. Being a supplier for both Burberry and Tom Ford, pioneers of the SNBN, Case A addressed that if orders are received duly and accurate, production could take place without delays. The shareholder of the Case A emphasized that upon the receipt of an order, they need three or four weeks to acquire the painted raw material, which is dependent on the size of the order and the kind of the material. Subsequently, cutting, production and packaging collectively take three months. 'We can do it in three months unless we are in a peak period.' Case A disclosed. In the case of a peak period three shifts are organized instead of a standard two. Case C further addressed that an iconic Italian luxury brand provides them with a sketch on Thursday night to be duly prepared for the runway show scheduled for the upcoming Tuesday. In this setting, Case C is asked to manufacture the apparel and deliver back to the brand after which the buying firm's own design team could finalize the look for the runway show. Given already strict timing, having a capacity to prepare an entire collection seems challenging not only for the manufacturers but also for the material suppliers. Flexibility is required to transform the sketch to an available product in a duly fashion.

Another big challenge is perceived as changing consumer demands. For example, 20 years ago buying firms and accordingly product manufacturers used to be demanding just four colours, namely black, blue, grey and brown, whereas recently about 15 colours are demanded per season, which requires the introduction of various yet smaller barrels for painting. Yet, painting within these smaller barrels led to a faster production. The production manager of Case D reveals that 'at the beginning, we deal with the same product and after our activities we obtain a semi finished product. This semi-finished product is painted with different colours. The phase of painting is the last one and we postpone it as much as possible in order to obtain more colours in a shorter time. This is the service for which we are recognized. Our clients know that if they need 10 colours in a week, we can do it.' Therefore, for the leather textile producers, having the order in advance and accuracy in terms of amount could be considered essential. Innovation capabilities emerge as significant prerequisites to enable the SNBN, e.g. new machine techniques to improve production quality and to enable production flexibility.

The competition pace was also announced. There appears to be a competitive driving force coming from the external pressures, implying that apparel manufacturers as well as textile producers feel the pressure of being left outside the game unless they accelerate their pace according Case D. Further, one of the biggest challenges is considered the human capital. Case C revealed that if all of their clients opt to implement the SNBN, at the current capacity the company might experience the 'mission impossible'. Thus, stronger relational capabilities are needed to better satisfy the bigger clients to 'keep' them. Therefore, manufacturing companies would need to improve their technical capabilities to meet the rapidly growing client needs. In this sense, Case C stressed that big buying firms cannot afford any mistakes, as timing is already limited. Thus, both textile producers and apparel manufacturers must maintain their performance at the best capacity. Case C also revealed that the design of fashion week seems a big challenge. All big buying firms are scheduled to display their collection at the same time, and therefore opting to implement the SNBN might not work for those who highly depend on outsourcing partners due to the fact that apparel manufacturers and textile producers might not meet the excessive orders coming from a number of big buying firms. It was further stressed that process optimization to reduce the production window of two months is not possible. Even the internal capabilities were improved and technical skills were developed, supply side would bring problems in delivering textiles and raw materials. Case B addressed that 'the SNBN would be more expensive. And also from the organizational point of view, the times get shorter while the production get more fragmented. The sizes of the lots would get shorter, yet with a higher quantity production costs much lower.' It was addressed that establishing long-term business relationships based on reliability and quality might earn a competitive a big advantage. Hence, being located close to the buying firm's headquarters (HQ) could be considered a big advantage. Short times and strict deadlines could be a competitive advantage for suppliers located in Europe in order to reduce leadtimes and increase delivery performance to implement the SNBN. Therefore, even though low-cost companies provide cheap prices, the SNBN seems not so adjustable for buying firms that opt to outsource from farther countries. Case C approaches the SNBN challenge as an opportunity. Case A correspondingly revealed that they do not see any barriers in their production performance, yet the SNBN will lead to divergence within the luxury industry. Even though, there is a belief that a consensus will be reached, at this very moment regarding the real consequences of the SNBN, no party in upstream SCs seems to be truly sure.

5 The Implications of the SNBN

- ***Large collections and high percentage of new items***. Every year luxury fashion brands are launching an impressive number of new items. Collections are getting wider and more frequent, and the percentage of carryover items seems to be shrinking. Tom Ford recently stressed that launching 16 collections per year for both Gucci and Yves Saint Laurent was creatively and personally exhausting, which resulted in him taking some time off prior to the launch of his eponymous label. Similarly, globally acclaimed designers, such as Stella McCartney and Vivienne Westwood, indicate that the connection between culture, customer base, and climate change for which each party in the business has a part to play. There appears to be the wider the collection the more difficult to implement SNBN.
- ***Flexible production networks***. Our research highlighted the extreme flexibility of some small Italian players which based their operations on a group of small artisan workshops located at short distance one to another (in the so-called artisanal—or industrial—districts: e.g. leather working district around Florence, shoe-making district in Parabiago, jewellery in Valenza, …). We were able to sort out several kinds of flexibility: the ability to produce different types of products, using different materials or with different finishes (product flexibility: essential when the artistic director introduces last minute a new and fancy material or process—e.g. Swarovski crystals on leather bags, wooden eyewear, plastic stilettos, …); the ability to produce small production batches with large width (many different models) and shallow depth (only a few pieces per model) (thus showing mix flexibility, as compared to those companies needing to plan large production runs of every model to stay efficient); the ability to doublefold (or more) production capacity in peak period, resorting to overtime, or working on weekends etc. (referred to as elasticity). Such flexibility would prove paramount in ensuring the speed of delivery in the few days preceding the fashion-shows.
- ***Near-shoring versus off-shoring***. And, of course, the closer the production is to the HQ the better (far away suppliers imply longer logistics leadtimes and poorer coordination—or at least longer communication processes—in the critical days).
- ***50 shades of Denim***. The wider the set of materials used by the creators, the more differentiated, the less commonalities (especially for new materials with high Minimum Order Quantities and long procurement leadtimes) the more difficult it would be to implement SNBN. Very often artistic directors are enticed by materials just launched by suppliers, and go for some last-minute tailoring to try and use said materials in one or more garments at the show (typically *just one*. As, in some case, it is seen as rather commonplace using the same materials in more than one item)—this is of course giving hard times to procurement and production.

- ***Directly Operated Stores***. Big players have a fairly large presence in the retail channel in terms of Directly Operated Stores (that's a plus as the SNBN part of the collection can be just shipped to the store avoiding show-rooms visits from buyers, order collections and the like) while small players distributed through independent retail will have a much larger need for coordination in the days before the catwalk and it seems that, in this case, SNBN could hardly work.

6 Discussion

Observation of the practices used by relevant fashion brands helps analyse why some companies cannot pursue the SNBN effectively, or furthermore what the drawbacks of the SNBN would be without modifying the approach to manage the SC structure starting from the development phases. Inevitably, the traditional approach to collection development and its relationship with the SC network Chain could be an obstacle to implement the SNBN. Conventionally, the "creative freedom" of designers—especially in less structured companies—gets to the point that all the supply system (from the product development function, to materials research, to sampling, to serial production) has to continuously adapt. Delivering the SNBN collection—even if this happens for the catwalk items—requires a very high degree of control over the development process so as to ensure that the timing of each activity is respected. This would imply the following points.

- First of all that the creation and production of a collection is the result of a strong collaboration among design, development and the SC; yet what we normally observe is that the SC is considered ancillary to the development, thus, at the end every need for compressing leadtimes and/or respecting target costs has to be fulfilled by the SC function. In order to achieve the SNBN approach, it is necessary to act as a system rather than separate functions.
- Secondly, the SC should be allowed to 'raise the hand' when some issues or difficulties emerge as a consequence of style choices. Indeed, no constraints should be given to the creative functions, but they have to be aware of all the impacts of their choices, including the selection of a material whose procurement leadtime is longer than the available time before the show.
- Finally, this means compressing leadtimes: Undeniably, the delivery in the stores has to coincide with the fashion show but the development process cannot be anticipated very much, as its early stages are driven by the timing of materials research and fairs. This means the introduction of specific methodologies (e.g.: lean techniques, agile philosophy) with the specific aim of reducing leadtimes: not only in the manufacturing, procurement and sampling processes, but also in the design phases. This would be a radical cultural change in the fashion world.

On the SC side, we can quote two further elements that facilitate the SNBN that are not common in fashion companies:

- *Forecasts*: the SNBN approach, which results in delivering in retail stores—physical or online—the same day of the show, requires a Make To Forecast production approach. Currently, most fashion companies still rely on a Buy To Order approach where the orders for materials are mostly made after collecting the sales orders. Probably, the diffusion of the SNBN approach will force fashion companies to set up explicit and effective demand forecast systems.
- *Capacity planning*: Delivering in short times and punctually, especially in the case of outsourcing production (which is the case of most of the fashion brands) requires being able to place production orders in periods when suppliers plants are unloaded. This requires a strong capacity-planning attitude, another rare characteristic of fashion companies. The difficulty is clearly higher for companies that make most of their business on the wholesale channel. Also small emerging designers that don't have high control and power over their supply system could be troubled by the diffusion of the SNBN choice.

7 Conclusion

It could be concluded that actually the SNBN is not completely new to the fashion world. There are companies that were able to make the product available very soon after its presentation on the catwalks or in the showrooms. In addition to the examples coming from the fast-fashion, the companies that are strongly based in internet and—as a consequence—used to satisfy customers' requests right after the publication of a new item on the web. The announcement of Burberry only gave visibility to this approach, in addition to legitimizing it as well as the irrationality of separating the presentation of men and women collections. The SNBN actually means reinterpreting both the essence of collection and the approach to physical distribution, this latter in coherence with an omnichannel view of the fashion world. There are basically three fundamentals for making the SNBN possible:

1. Time compression along the whole set of processes: The critical path for development and production has to be shorter. Because what is presented and immediately sold is not a self-standing collection; it rather offers an initial taste of the main collection that will be in stores some months later, hence they have the same design origin.
2. Proximity sourcing: Reshoring is a fundamental issue in fashion, especially for brands which are sensitive to the 'made in' critical success factor. However, among the major reasons why it is convenient to manufacture close to the destination markets there is a necessity of speeding up deliveries. If the product is already close to the market—either in the warehouse or as a work-in-process—

this becomes a condition for making possible the immediate delivery right after the presentation on catwalks or showrooms.

3. Optimization of sourcing and production: The adaptation of lean techniques to the production floors can improve the time performance of about 30% and take leadtime down to two weeks. This means that, if the sales campaign lasts 3 weeks, it is possible to launch production right at the beginning of the campaign.

References

Bobila M (2015) The CFDA is rethinking the fashion week format. Fashioista. http://fashionista.com/2015/12/cfda-studies-fashion-week. Accessed 12 Feb 2016

Fernandez C (2016) Alexander McQueen will return to London for fall 2016 runway show. Fashioista. http://fashionista.com/2015/12/alexander-mcqueen-london. Accessed 10 Apr 2016

Gonzales E (2016) Why you won't find see now, buy now shows at Paris Fashion Week. Harper's Bazaar. http://www.harpersbazaar.com/fashion/designers/a14334/see-now-buy-now-paris-fashion-week/. Accessed 9 Apr 2016

Hoang L-M (2016) Burberry sparks "see now, buy now" debate at London Fashion Week. Reuters. http://www.reuters.com/article/us-fashion-london-idUSKCN0VW171. Accessed 11 Apr 2016

Lockwood L (2015) NYFW going consumer? WWD. http://wwd.com/fashion-news/designer-luxury/cfda-boston-consulting-fashion-shows-consumer-10297602/. Accessed 11 Mar 2016

Maddeaux S (2016) The see-now-buy-now model could ruin indie designers. Fashion magazine. http://www.fashionmagazine.com/fashion/2016/02/08/the-see-now-buy-now-model-could-ruin-indie-designers/. Accessed 8 Apr 2016

Mau D (2016) Bill Blass is back. Fashioista. http://fashionista.com/2015/10/bill-blass-chris-benz-relaunch. Accessed 6 Apr 2016

Mower S (2016) Burberry and Tom Ford both announce see-now-buy-immediately collections for September. Vogue. http://www.vogue.com/13395498/burberry-tom-ford-announce-see-now-buy-immediately-collections/. Accessed 19 Feb 2016

Wischhover C (2016) How will "see now, buy now" fashion shows affect beauty brands? Fashioista. http://fashionista.com/2016/02/new-runway-show-format-beauty-sales. Accessed 1 Mar 2016

Yin RK (1994) Case study research: design and methods. Sage Publications, thousand oaks, CA

Author Biographies

Alessandro Brun is an associate professor of quality management at Politecnico di Milano. His research interests cover application of Supply Chain Management, Quality Management and Operations Improvement methodologies in high value added sectors, with special focus on Luxury Management. With more than 100 papers published on scientific journals and international conferences on the above topics, is one of the most prolific authors in the worldwide scientific community on the specific "Luxury Supply Chain". In 2012, he created and launched the "International Workshop on Luxury Retail, Operations and Supply Chain Management"—the first international conference on Luxury Supply Chain.

Cecilia Maria Castelli got a Ph.D. at Politecnico di Milano with a thesis focused on Managing Supply Chains in the Luxury Market. She works as business consultant in the Global Fashion Unit of The European House—Ambrosetti, where she is involved in projects dealing mainly with supply chain, operations and industrial plans for major fashion-luxury brands. Her areas of interest do not limit to the luxury-fashion industry but spans over other industrial/service sectors. She collaborates with the School of Management of Politecnico di Milano, teaching Luxury Supply Chains and Luxury Retail Management. She is part of the Organizing Committee of the "International Workshop on Luxury Retail, Operations and Supply Chain Management".

Hakan Karaosman is a sustainability management professional and a Ph.D. researcher with international experience in sustainability reporting, business improvement, and sustainability strategy. Previously, he worked as a consultant to companies in various industries for their Carbon Disclosure Project (CDP), and Corporate Social Responsibility (CSR) reports. He holds B.Sc. in Environmental Engineering, MBA, and M.Sc. in Management in Engineering of Energy and Environment. He is currently pursuing his Ph.D. studies in European Doctorate in Industrial Management, a double degree program funded by the EU, and hosted by KTH (Sweden), POLIMI (Italy), and UPM (Spain). His driven focus is towards sustainability integration in fashion supply chains.

Development of a Fashion Buying Education Program for an Apparel Retail Company

Sedat Sezer, Burcu Guven, Emel O. Karaoglu, Cevza Candan, Zafer Onder, Emel Aydemir and Gulperi Tandar

Abstract In order to stay ahead of the marketplace, today's retailers need to be smarter in managing and controlling what goes on their shelves. Therefore, it is essential for professional buyers to focus on their product selections, and to ensure that these selections are the products which will satisfy their customers. In other words, professional retail buying is all about being able to anticipate consumer demands and to create ideas that can be converted into exciting products that fulfil consumer's desires. However, this is not always an easy task to do for especially Turkish retail buyers as there is a lack of professional buying courses applicable for Turkish apparel retailers. The courses available are generally too theoretical and/or too general for practical implementation. As a result of that, LC Waikiki Corporate Academy, in collaboration with Faculty of Textile Technologies and Design, Istanbul Technical University, aimed to establish "an in-house buying certificate program" which covered the how-to, the what-to and the when-to, with easy to understand material and case studies that would enhance their buyers' skills and facilitate the management of their daily activities. The program also aimed to help

S. Sezer (✉) · B. Guven · Z. Onder · E. Aydemir · G. Tandar
LC Waikiki Corporate Academy, İstanbul, Turkey
e-mail: SEDAT.SEZER@Lcwakiki.com

B. Guven
e-mail: BURCU.GUVEN@Lcwakiki.com

Z. Onder
e-mail: ZAFER.ONDER@Lcwakiki.com

E. Aydemir
e-mail: EMEL.AYDEMI@Lcwakiki.com

G. Tandar
e-mail: GULPERI.TANDAR@Lcwakiki.com

E.O. Karaoglu · C. Candan
Faculty of Textile Technologies and Design, Istanbul Technical University, İstanbul, Turkey
e-mail: onderem@itu.edu.tr

C. Candan
e-mail: candance@itu.edu.tr

R. Rinaldi and R. Bandinelli (eds.), *Business Models and ICT Technologies for the Fashion Supply Chain*, Lecture Notes in Electrical Engineering 413,
DOI 10.1007/978-3-319-48511-9_5

participants to acquire knowledge of the discipline of fashion buying, to explore current global and local issues in retail buying, and finally to enhance the development of participants' global awareness, critical and analytical thinking and communication skills.

Keywords Apparel · Fashion buyer · Retail buying · Professional education

1 Introduction

Looking at the basic structure of the fashion industry until the late 1980s, traditionally fashion apparel retailers used their capability of forecasting consumer demand and fashion trends long before the actual time of consumption in order to compete in the market (Guercini 2001). However, recent years have seen fashion retailers compete with each other by ensuring speed to market with their ability to provide rapidly the fashion trends revealed by fashion shows and runways. According to Taplin (1999), such retailers could be credited with the adoption of 'quick fashion' that is an outcome of an unplanned process on the reduced time gap between designing and consumption on a seasonal basis. With the emergence of small collections of merchandise, fashion retailers are encouraging consumers to visit their stores more frequently. This indicates a shorter life cycle and higher profit margins from the sale of fast selling merchandise, skipping the markdown process altogether. In addition, consumers are fashion conscious, and at the same time expect affordable prices as well (Gabrielli et al. 2012). Literature on fast fashion implies that rapid responsiveness techniques such as just-in time, quick response, and agile supply chains can be valuable to the fashion industry because such techniques can create a competitive edge in the market (Bruce et al. 2004; Christopher et al. 2004; Fiorito et al. 1995; Sohal et al. 1998; Bhardwaj and Fairhurst 2010). The dramatic change in the fashion apparel industry has forced the industry as well as academia to broaden, redesign and align their research to match the needs of fashion markets including fashion consumers, in the twenty-first century.

2 Literature Survey

Regarding the aforementioned changing dynamics of fashion retail industry, the studies on organisational buyer behaviour theory have become one of the widely discussed topics in the literature of retailing and marketing. Though the traditional models of this theory can be applied to the retail buying process (Sheth 1973, 1981; Webster and Wind 1972), they were developed decades ago and therefore these models do not reflect contemporary changes- significantly influenced by recent developments in computer technology and globalisation- in retail buying practice

(Fiorito et al. 2010; Johansson 2001; Schröder 2015). Also, they are specifically on purchase decision-making, which forms a minor proportion of the role of a buyer (Swindley 1992).

The retail buying literature focuses primarily on studies of department stores or retail buying groups, whereas retail marketing texts usually cover consumer buying theory in depth, rather than referring the roles of retail buyers. In the early 1990s, the retail buying role was investigated by Swindley (1992) via a survey of buyers in the UK and by Wall et al. (1994) who conducted a qualitative study in Canada. Swindley's survey of 63 respondents included 25 fashion buyers and found that the buyer's role was not restricted to the activities of product selection and supplier liaison, though these were generally their main responsibilities (Swindley 1992). Other key activities for buyers in the study were monitoring product performance and profitability. Wall et al.s study contained interviews with buyers for clothing retail chains and garment suppliers, and a key finding was that product sourcing was the most complex aspect of the retail buying role (Wall et al. 1994). McGoldrick, however, offered an outline of the buyer's role, by stating that "supplier selection, appraisal and negotiation are core responsibilities of retail buying departments, and their most time-consuming activity" (McGoldrick 2002). Jacobsen's description of the retail buyer's role (Jacobsen 2009) was orientated towards managing stock, budgets and profit margins, whilst (Cash et al. 2006) stated that retail buyers must have engaged in the following activities: Consumer research; Selecting products to meet consumers' needs, deciding on quantities and delivery times; Sourcing suitable suppliers; Planning logistics from the order to the final sale; Working with other departments to promote products; Managing records for the product range to assess profitability; and Observing competitors'strategies. Fiorito and Gable (2012) summarised the buying role as follows: planning; purchasing; assorting; controlling and communicating. Fernie and Moore (2003) described the principal activities in retail buying rather differently, though they remained compatible with Fiorito and Gable's summary (Fernie and Moore 2003) which was as follows: Analysis of market opportunity; Creation of the merchandise plan; Selecting the supply base; Product development and supplier performance management; and Presentation of merchandise at point-of-sale. Furthermore, Johansson and Burt (2004) conducted a comparative study of buying for private and manufacturer brands in the UK, Sweden and Italy, concluding that involvement in private brands meant that the buying process would be more extended, in terms of the tasks and activities performed and managed, and therefore more complex than that for manufacturer brands. Goworek (2014) investigated own-label buyers' roles and responsibilities within the context of fashion retailing in the UK, as distinct from the roles of buyers of branded merchandise supplied by other companies. The research concluded that buying own-label fashion products was largely similar to buying branded merchandise, with one of the main exceptions being in the area of new product development. It also stated that retail merchandisers were found to play a significant role within the buying function, yet there was a relative lack of collaboration between buyers and marketers within the sample.

Literature survey on retail buying further showed that the most important part of a buyer's role in a retail company was satisfying company objectives by making accurate and timely decisions of merchandise planning and assortment planning. This is mainly because decisions related to the acquisition of merchandise are critical to the profit potential of a retail company. The creative side of a buyer is the ability to understand customer, spot trends, and use intuition to choose merchandise with terrific sell through. The analytical side of a buyer, however, is the ability to evaluate merchandise and judge whether the selected products and the quantity of the products are suitable for their target customer. To succeed in this career, buyers need to have foresight and develop skills in people as well as time management. It is very difficult to ascertain solely from CVs and interviews whether or not a person has the right qualities to be a buyer, as most of these will only be developed by exposure to the fashion buying environment. Even with extensive skills and experience, a buyer who is new to a company will require a certain period of training and readjustment to become familiar with different systems and terminology. Companies that do not recognise and plan for this factor could lower their profits as a result when the new buyer's range is launched. In summary, it appears that retail buyers depend on more qualitative information sources (e.g., own knowledge, peer opinion, consumer publications) than quantitative information sources (e.g., past sales record) in a product-specific, decision-making process. Moreover, they may face difficulty in interpreting a qualitative decision, resulting from qualitative information sources, into a quantitative demand forecast for the first order (Zentes et al. 2011; Levy et al. 2012; Schröder 2015; Rabolt and Miler 2009).

Due to the reasons outlined above, retail buying education is of particular importance especially for Turkish retail apparel companies, as international/national professional buying courses have limited applicability to the Turkish apparel retail companies, since they are either too theoretical or too general for practical implementation. In order to fill the gap, it was aimed to develop an in-house buying certificate program, which sets an educational standard to improve and to complete the required competencies for the buying professionals of various levels (i.e. junior, mid or senior). In doing so, own-label buyers' roles and responsibilities within the context of fashion retailing both in Turkey and in international markets were examined. Also, contemporary retail buying processes in national/international fashion markets were investigated so that it could be possible to establish and compare the responsibilities of buyers within an apparel retail company as well as the interaction between buying and other retail functions in the company.

3 Methodology

The Buying Education and Professional Development Program is an in-house buying certificate program which sets an educational standard to develop and to complete the required competencies for the Buying Professionals working for the company at either junior, mid or senior levels. The buying professionals (or buyers) are the white collar personnel having different backgrounds such as textile engineers, industrial engineers, textile or fashion designers, etc., and also having different experiences in apparel retail sector as being a merchandiser, garment technologist, associate or assistant buyer or allocator etc. Thus, everyone (or someone) in the organization can be assigned to a buyer position after he or she has spent a considerable time in various positions and has learnt the environment, dynamics and operations in the apparel retail business in general. However, the buyer's task involves in much more complex duties and responsibilities ranging from keeping abreast of current market trends and economic conditions and understanding the consumer through collaborating with the planners and designers in developing merchandising strategies for a product line and planning and selecting the merchandise assortments to negotiating with suppliers, procuring merchandise and helping the coordination of promotion activities. In changing and extending role of the retail buyer, leadership, broad vision, goal–setting ability, merchandising knowledge and skills, analytical ability, human relation and communication skills are the major qualifications needed. Therefore, the program developed aims to prepare buyers to attain a number of qualifications by the time of graduation, described as the program outcomes (POs) for the professional practices of buying. The program outcomes relate to the skills, knowledge, and behaviours that participants acquire as they progress through the program. Buyers are expected to develop an expertise in the field within a few years of graduation, described as program educational objectives (PEOs). In our outcome–driven approach, we first defined the PEOs by reviewing the buyer's role in the apparel retail sector as well as in the organization while regarding the company's policies and values (http://www.tgsd.org.tr/sektorel-bilgiler/raporlar, http://www.ihkib.org.tr/tr/bilgi-bankasi/raporlar/aylik-ihracat-bilgi-notlari/k-294, http://corporate.lcwaikiki.com/MainPage). The second step of the study was to describe the program outcomes (POs). In doing so, the study adopted a case study approach, including in-depth secondary data analysis on retail buying, fashion retailing, contemporary buying processes, and international fashion buying education in general; and focus groups with buying personnel from different levels (junior, manager, etc.) to be able to define PEOs and POs in a measurable/assessable way, which was supported by documentation and observation (Cooper and Schindler 2014; Quinlan 2011) to acquire comparative information about their products, services, etc. Accordingly, PEOs of the Buying Education and Professional Development Program are given as follows:

1. Graduates will demonstrate a proficiency while managing the buying processes, be able to identify and solve the arisen field problems in context of their duties and responsibilities.
2. Graduates will demonstrate a collaborative approach in the organization to conduct multidisciplinary apparel retail operations in order to accomplish the company's set out targets.
3. Graduates will concern with the environmental, health and ethical issues of buying processes.

4. Graduates will pursue the continuous self-improvement, be able to initiate, modify and/or redesign new buying processes and/or research and development activities to fulfil the specific objectives or changing expectations.

A critical focus of the education within LC Waikiki Corporate Academy was to afford the white collars of varying backgrounds and abilities every opportunity for achieving success in "buyer" profession. The Academy, in collaboration with ITU, systematically studied various international fashion buying education programs which were run by different institutions/universities all over the world, together with the literature on fashion as well as retail buying in order to accurately define the skills and level of professional knowledge expected from a buyer performing in fashion retailing arena. In parallel with the literature survey, the feedback input was also sought from the Company policies and values, the senior buying teams of the Company, and from the Turkish fashion/apparel retail industry (http://www.tgsd.org.tr/sektorel-bilgiler/raporlar, http://www.ihkib.org.tr/tr/bilgi-bankasi/raporlar/aylik-ihracat-bilgi-notlari/k-294, http://corporate.lcwaikiki.com/MainPage). Accordingly, the program outcomes from 'a to k' for the Buying Education and Professional Development Program, which stated the competencies expected from buyers, were defined. Furthermore, each outcome was broken-down into the corresponding performance criteria in order to set the Company's accreditation standards for these competencies in terms of level of professional knowledge, skills and responsibilities. Abbreviated statements of these program outcomes are given as follows:

(a) Competence in consumer, market and trend analysis to develop commercial intuition and foresight.
(b) Competence in collection (range) development and its implementation.
(c) Professional and ethical responsibility.
(d) Fundamental knowledge both on merchandise planning and its implementation to collaborate with planners.
(e) Leadership in procurement and production management processes.
(f) Understanding of the marketing principles to collaborate with advertising and promotion people.
(g) Competence in analysis and interpretation of both internal and external data to make decisions on sale and profit maximization.
(h) Effective verbal and written communication skills.
(i) Functioning on different teams.
(j) Competency of information technologies (IT) and computer use related to buying practices.
(k) Knowledge and understanding of contemporary issues influencing fashion retail practices and complementing business life.

Taking the POEs and program outcomes into consideration, the curriculum for Buying Education and Professional Development Program was developed such that it had its own academic calendar of 9 months. It was composed of 19 courses (compulsory and elective ones), 34 credits. Problem based learning approach was adopted for the program so as to facilitate problem solving skills and analytical ability of the participant buyers. Participants were required to do projects and case studies to help them integrate knowledge of the industry. Lectures were structured to convey theories and concepts for the discipline of fashion retail buying. E-learning videos for some courses were designed to supplement lectures to facilitate learning. The combined approach of the in-house instructors with the university faculties was used to develop critical thinking ability of the participants towards practical issues of fashion retail buying. The Program had also a detailed assessment process (see Sect. 3.3) which could make participants being evaluated individually by continuous coursework and homework.

3.1 Continuous Improvement

The methodology adopted for designing the program was such that the "continuous quality improvement loops" similar to the ABET accreditation approach were formed to fully describe the strategy and practices involved in developing and implementing and program learning outcome-driven continuous improvement system (http://www.abet.org/, http://www.iso.org/iso/home.htm). The Quality Assurance System included an assessment program with three standard processes: the course assessment process—Course Loop, the program outcomes assessment process—Program Loop and the constituencies' assessment process- Constituencies' Loop (see Fig. 1).

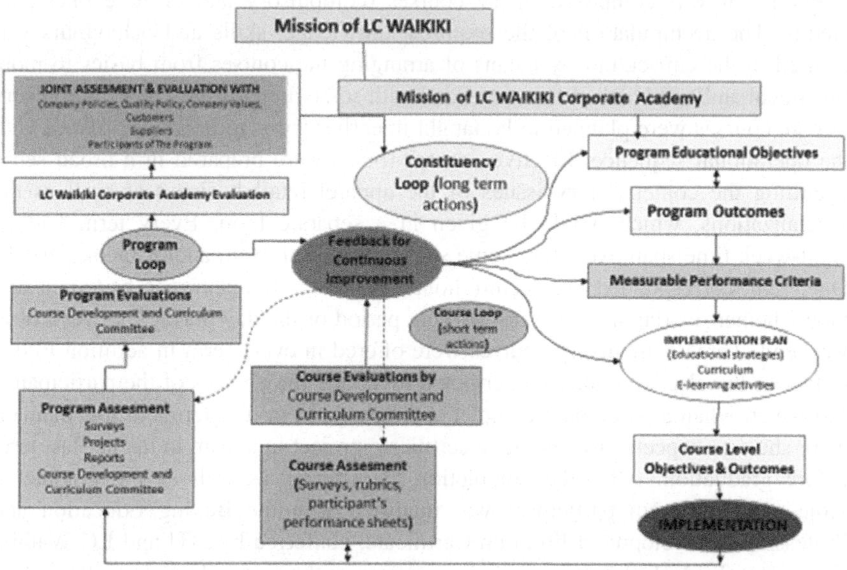

Fig. 1 Quality assurance system for the buying education and professional development program

Each process supported and complemented the other, and all three processes were producing tangible results. The focus of the assessment program was on participant learning, and on how the program could facilitate the participant learning more effectively. A key component of this model was the identification of constituency needs, the tools to address those needs, and a timetable for collecting and analysing data, and for disseminating results for program planning.

3.2 Curriculum and Its Implementation

The curriculum consisted of courses with syllabi which, when taken as a whole, met the Buying Education and Professional Development Program Objectives. Course-level objectives, strategies, and outcomes helped to define what learning outcomes were expected as a result of a specific course. Thus, the courses in the curriculum were structured to convey the basics and functions of fashion buying profession, while considering the managerial and general education contexts as well, all constituted the theories and concepts for the discipline of fashion buying as a whole. Each course content was deliberated to satisfy the predetermined course level learning outcomes within either a single context or a combination of them, say 50% basic and 50% functional education or a 100% managerial education, etc. The course plan in syllabus covered topics distributed over the weeks so that each topic indicated its association with one or more learning outcomes of a particular course. Moreover, the level of course contribution (little, partial or full) to every single program outcome was well defined. The program had its own academic calendar of 9 months. It was composed of 19 courses (compulsory and elective ones), 34 credits. The accumulation of the required knowledge, skills and behaviours was assured in the curriculum by means of arranging the courses from basics to more functional and/or advanced ones provided with some prerequisites. The compulsory or core courses were planned to be taught in earlier terms to be in accordance with the operational sequence. Elective courses (pool) were prepared in a broad sense regarding the contemporary issues of the apparel retail business as well as the specializations, which would be given in a separate term. Every term had an eight-week time span excluding some subsequent final exam and/or break periods. One credit corresponded to 1 lecture hour. Courses were planned to be taught in a time whether overlapping the late working period of the day during the week or in weekend. Three compulsory courses were offered in every term in addition to one or two of the elective courses determined by taking the choices of the participants. Moreover, when a participant completed the program in four terms in this manner, he or she was expected to conduct a certificate project in a team in his/her last term before graduation. After the completion the courses as well as the graduation project, the program participant was qualified with the Buying Education and Professional Development Program Certificate, conferred by ITU and LC Waikiki Corporate. Table 1 shows one of the courses in the curriculum as well as their contributions to the program outcomes in order to exemplify the process.

Table 1 Buying education and professional development program outcomes—curriculum contribution matrix (contribution of the course: 1. little, 2. partial, 3. full)

Term	Code	Course name	Credit hours*	Outcomes of buying education and professional development program										
				a	b	c	d	e	f	g	h	i	j	k
I	BUY101	Principles of apparel retail supply chain management	2	2				2		2			1	2

A problem based learning approach was adopted for the program so as to facilitate problem solving skills and analytical ability of the participant buyers. They were required to do projects and case studies to help them integrate knowledge of the industry. Lectures were structured to convey both theories and practices of fashion buying. The combined approach of the in-house instructors with the university faculties was used to develop critical thinking ability of the participants towards practical issues of fashion buying. E-learning videos for some courses were designed to supplement lectures to facilitate learning. Based on interactive teaching and learning approach, participants were aimed to be assessed individually by continuous coursework and homework.

3.3 Assessment

The course assessment process was developed to ensure the educational competencies of the individual courses. Each had objectives, outcomes and a defined relationship with the part of the curriculum to be mastered, specified by the LC Waikiki and ITU training team member(s). Course-level objectives, strategies, and outcomes helped to define what learning outcomes were expected as a result of a specific course.

In the program loop, several independent assessment strategies were established to ensure that the program learning outcomes were covered and accomplished by the program as a whole (see Table 2). The program assessment data, say exit surveys, exit interviews, and graduation project survey results were collected, evaluated and reviewed once the program was completed.

Table 2 indicated both course tutorial (lecture) hours as well as the required extra study hours during the term for each course. Each row for a course in Table 2 also illustrated the distribution of the course activities in detail (i.e. homework & projects, computer use, laboratory work, other activities), in terms of their quantities, loads and weights on the course success in percent. The program altogether offered 272 tutorial hours, and the participants were expected to study additional 177 h for the course activities which totally correspond to 450 h to take the certificate. For establishing, revising and achievement of the "Buying Education and Professional Development Program" outcomes, the roles played by the Academy

Table 2 Assessment strategies for the courses

Term	Code	Course name	Credits	Program assessment tools											
				Course hours	Extra curricular activities (hours)	Total hours	Midterm exam	Homework	Term paper	Project	Computer applications	Lab. applications	Self-evaluation repor	Quiz	Final exam
I	BUY102	Apparel production and quality management	2	16	10	26	–	–	1 (30%) 5 h	–	–	–	1	–	1 (70%) 5 h
		Total	–	–	–	–									–

Table 3 The main roles of LC waikiki corporate academy and LC waikiki-ITU for continuous improvement process

LC waikiki corporate academy	• To monitor the run of the program • To continuously improve the e-learning tools available and develop new ones if necessary • To manage the program evaluation activities and collect data according to these activities at the end of each term (module) • To administrate course/instructor opinion surveys, collect data and evaluate the results • To interview with participants, and review individual proposals of participants for the ongoing progress of the program • To evaluate implications related to the general performance of the program • To review the changes in curriculum and new course proposals • To take the necessary actions and corrections whenever necessary
Course development committee (LC waikiki—ITU)	• To address short-term curriculum issues and ensure the ongoing evaluation and progress of the program • To monitor courses and prepare reports for each course including recommendations for the actions and corrections at the end of the semester report the curriculum committee and the department • To monitor long term trends in buying education • To recommend strategic curricular adjustments to the program (new courses and individual course objectives or additions, deletions and changes in syllabi, and/or major program changes)

with respect to course creation, modification, and evaluation were, however, listed in Table 3.

Table 4 shows "Program Outcome Assessment Matrix" for multiple assessment tools regarding both assessment period and the power of indicatiors. In course-level assessments, course and instructor opinion surveys were conducted every term following the final exams. The surveys were designed to include both multiple choice and open-end questions. Participants were asked to assess each course learning outcome and the course's instructor(s) based on 7 Likert scale in order to obtain feedback reflecting relevant learnings for any specific course and the teaching effectiveness of the instructor(s). If 70% of the responses indicated a satisfaction level above 4 (or more) based on 7, the learning outcome was assumed to be achieved. Instructors also discussed the results of 'course and instructor opinion surveys' for each course taught by them, and reported their opinions and considerations for the course.

Several direct assessment tools such as midterm exams, quizzes, homework, final exam, etc., were used to evaluate the quality and contents of the students' work executed. In addition to the multiple course assessment tools used to evaluate the

Table 4 Program outcome assessment matrix

Program outcomes assessment tools (outcome indicators)	(a) Competence in consumer, market and trend analysis to develop commercial intuition and foresight	(b) Competence in collection (range) development and its implementation	(c) Professional and ethical responsibility	(d) Fundamental knowledge both on merchandise planning and its implementation to collaborate with planners	(e) Leadership in procurement and production management processes	(f) Understanding of the marketing principles to collaborate with advertising and promotion people	(g) Competence in analysis and interpretation of both internal and external data to make decisions on sale and profit maximization	(h) Effective verbal and written communication skills	(i) Functioning on different teams	(j) Competency of information technologies (IT) and computer use related to buying practices	(k) Knowledge and understanding of contemporary issues influencing fashion retail practices and complementing✚✚✚✚✚ businesslife
A. During education period											
Course level assessments defined in each course syllabi (midterm exam, homework, term paper, project, oral presentation, quizzes, computer labs, practices, self-reporting, final exam)	✚	✚	✚	✚	✚	✚	✚	✚		✚	✚
Performance appraisals (evaluations for the results of course learning outcomes)	✚	✚	✚	✚	✚	✚	✚	✚		✚	✚
Rubrics								✚	✚		
Course opinion surveys	✓	✓	✓	✓	✓	✓	✓	✓	✓	✓	✓
Instructor course evaluation report	✓	✓	✓	✓	✓	✓	✓	✓	✓	✓	✓
Buying certificate project and survey	✚	✚	✓	✚	✚	✓	✚	✚	✚	✓	✚
	✓	✓	✓	✓	✓	✓	✓	✓	✓	✓	✓

(continued)

Table 4 (continued)

Program outcomes assessment tools (outcome indicators)	(a) Competence in consumer, market and trend analysis to develop commercial intuition and foresight	(b) Competence in collection (range) development and its implementation	(c) Professional and ethical responsibility	(d) Fundamental knowledge both on merchandise planning and its implementation to collaborate with planners	(e) Leadership in procurement and production management processes	(f) Understanding of the marketing principles to collaborate with advertising and promotion people	(g) Competence in analysis and interpretation of both internal and external data to make decisions on sale and profit maximization	(h) Effective verbal and written communication skills	(i) Functioning on different teams	(j) Competency of information technologies (IT) and computer use related to buying practices	(k) Knowledge and understanding of contemporary issues influencing fashion retail practices and complementing✚✚✚✚✚ businesslife
Monitoring course grade point achievement distribution											
Monitoring participant transcript	✓	✓	✓	✓	✓	✓	✓	✓	✓	✓	✓
Course Development and curriculum committee report	✓	✓	✓	✓	✓	✓	✓	✓	✓	✓	✓
Instructor opinion surveys	It is a measure of general success of instructors										
B. Before and after education period											
Program exit survey and interview	✚	✚	✚	✚	✚	✚	✚	✚	✚	✚	✚
Graduate evaluations	✓	✓	✓	✓	✓	✓	✓	✓	✓	✓	✓
Manager evaluations	✓	✓	✓	✓	✓	✓	✓	✓	✓	✓	✓
LC waikiki corporate academy evaluation	✓	✓	✓	✓	✓	✓	✓	✓	✓	✓	✓
	It is a measure of general success of education										

(continued)

Table 4 (continued)

Program outcomes assessment tools (outcome indicators)	(a) Competence in consumer, market and trend *analysis to* develop commercial intuition and foresight	(b) Competence in collection (range) development *and its* implementation	(c) Professional and ethical responsibility	(d) Fundamental knowledge both on merchandise planning and its *implementation* to collaborate with planners	(e) Leadership in procurement and production management processes	(f) Understanding of the marketing principles to collaborate with advertising and promotion people	(g) Competence in analysis and interpretation of both internal and external data to make decisions on sale and profit maximization	(h) Effective verbal and written communication skills	(i) Functioning on different teams	(j) Competency of information technologies *(IT) and* computer use related to buying practices	(k) Knowledge and understanding of contemporary issues influencing fashion retail practices and *complementing*✚✚✚✚✚ businesslife
Demand for the program											
Job placement data to buyer positions											It is a measure of general success of education.
Budget and investments for the program progress	It is a measure of general success of education.										

✚: Direct (substantial) indicator; ✓: Indirect (slight) indicator

course achievements, rubrics were also implemented to assess both communication and teamwork skills of the participants. At the end of each term, Course Development & Curriculum Committee reviewed the assessment results and prepared a report for each of the courses, including particular recommendations on the actions to be taken and on some corrections, and submits to the LC Waikiki Corporate Academy.

The program assessment data, exit surveys, exit interviews, and graduation project survey results were collected, evaluated, reviewed and finally reported to the relevant units defined in the organizational chart of the Company, once the program was completed. By doing so, it was aimed at achieving the consistency and quality of the courses taught based on the feedback of the success of the course learning outcomes.

In the program exit survey, recent graduates were asked whether they feel that the program outcomes have been adequately covered by the cumulative program curriculum, and to rate the quality of the preparation in both professional and technical learning outcome areas with multiple choice and open-end questions. In addition, LC Waikiki Corporate Academy interviewed with a group of exits on the subjects, which contributed to program continuous improvement.

Graduate evaluations were planned to be held every two years as an assessment strategy to gather feedback from the graduate buyers about the gained competencies. Manager evaluation was another feedback for the buyer's gained competencies. Demand for the program and the job placement data were important indicators for the general success of the program.

4 IT Facilities and E-learnings at LC Waikiki

Table 5 shows how the program outcome (j) is converted into three performance criteria as well as the course level learning outcomes for four different courses correlated with this outcome. The IT facilities and e-learnings were included by the program in this relation.

4.1 E-learning Applications for Courses

BUY203: Retail Mathematics and Introduction Planning: This course connects an "e-learning application" to a "table workshop". A video is structured to teach or to remind the basic elements and terms of retail mathematics, e.g. retail price, gross margin, mark-up definitions and calculations, discounts, turnover, stock-to-sales ratios, rate of sales (ROS), sales per square meter, etc. After the participants will have learnt the given content, they will play a table game during the lecture time, in which simulated store sales will be managed individually by deciding on price and by making the required retail calculations to achieve a certain profit target.

Table 5 Program Outcome (j)- Competency of information technologies (IT) and computer use related to buying practices

COURSES contributing to the program outcome (j)	Performance criteria of the program outcome (j)			
	J1. Knowledge on IT and ability of computer use for MS office applications, at the fundamental level	J2. Knowledge and employment of in-house software systems related to buying practices	J3. Knowledge and employment of online research and information services related to buying practices	
BUY 101 Principles of apparel retail supply chain management	II. Knows the principles of apparel retail supply chain, its operations and information technologies, planning, and management			Course learning outcomes for the participant
BUY201 fashion market research and forecasting			IV. Knows and employs the tools available for the fashion forecasting and trend analysis	
BUY202 garment design management		V. Knows the software used in product development and pattern making		
BUY301 management information systems and data management	I. Knows basic technologies, concepts and processes of information systems II. Understands the functions of information systems for organizations	III. Be familiar (knows) about in-house information systems and data bases IV. Be able to collect and analyse sales data using in-house information systems and data bases VI. Be able to follow the product related processes using in-house information management as well as database systems	V. Be able to collect and analyse consumer and market research data using in-house information systems and data bases	

BUY403: Business Ethics: So far as this course is concerned, there is a comprehensive e-learning about the ISO27001 Information Security and Security Infrastructure, which comprises the subjects on what to and how to the information security is ensured/managed and explains the related responsibilities.

4.2 IT Applications for Courses

BUY 101: Principles of Apparel Retail Supply Chain Management: Participants will be taught in this course to gain a familiarity for the information technologies (software, simulations, etc.) used in the organisation for the supply chain.

BUY202: Garment Design Management: Participants are informed about various types of softwares used in product development and pattern making processes, such as Gerber (AccuMark) Pattern Maker and 3D (AccuMark) V-Stitcher™. The buyers are expected to have managing abilities in the product development processes, be able to decide on fit for the products and silhouettes, as well as to analyse feasibility and producibility of the product. Therefore, these softwares will be taught in this course to develop an understanding of participants about the programs' functions and practices.

BUY301: Management Information Systems and Data management: This course aims to teach the basic technologies, concepts and components of management information systems and to provide an analytical and managerial perspective for benefiting from information systems in solving organizational/managerial problems. The following information systems and data bases used in LC Waikiki will be overviewed in the context of the course:

FlexPLM *is used in the organisation to manage the processes from the stage of product design to that of pricing for ordering. It provides line planning, specification management, merchandising and other essential PLM capabilities for managing the company's complete assortment of products* (http://www.ptc.com/product-lifecycle-management/windchill/flexplm).

TEMATEKSTIL *is a program uniquely developed for the organisation. This in-house software is an information and data management system in which products are specified, orders are placed, productions, orders and stock levels are followed.*

TEMANUMUNE *is another software used in the preparation of the technical documents of the product designs needed for the production and order placements (which is going to be replaced by the employment of FlexPLM).*

The basic use of ***MS OFFICE Programs****, particularly Excel will also be improved by practical work.*

Tema Web Portal *is an information share platform established based on Microsoft Sharepoint system and actively used by all departments in the organisation.*

WGSN *is the professional fashion blog which is the most comprehensive well-known the world over. WGSN is the world's leading trend authority. The services* *cover fashion and lifestyle forecasting, data analytics, crowd-sourced design validation and expert consulting* (http://www.wgsn.com/en/wgsn/).

BUY503S: Visual Merchandising: Polytropon Mock shop is a set-up of virtual reality and visual reporting tools that helps participant to build fully merchandised interactive 3D stores, to automatically generate planograms which provide a complete visual guide to every fixture in the store, to visually analyse the product data given, and to create storyboards and range books in minutes (http://www.polytropon.com/en/Software/MockShop-Virtual-Merchandising.html). Mock shop applications will be practiced in this elective course.

5 Conclusions

The project discussed in this paper was the unique one so far as the Turkish Apparel Industry is concerned. This was mainly because of the fact that unlike some other certificate programs run by private institutions, it was a truly tailor-made program in accordance with the company as well as the Turkish industry dynamics. In doing so, the skills and abilities required to succeed as a professional buyer had been discussed and analysed by studying the literature, the similar international programs available, and the needs/expectations of the company/the Industry defined, in general. And then, the buying certificate program was structured such that it was composed of two levels: the first level was for senior buyers whilst the second (advanced) level was meant for buying managers. The first assessment results of the program revealed that Buying Education Program was implemented in such a way that it successfully satisfied the expectations of the stakeholders.

References

Bhardwaj V, Fairhurst A (2010) Fast fashion: response to changes in the fashion industry. Int Rev Retail Distrib Consum Res 20(1):165–173

Bruce M, Daly L, Towers N (2004) Lean or agile: a solution of supply chain management in the textiles and clothing industry? Int J Oper Prod Manage 24(2):151–170

Cash RP, Thomas C, Wingate JW, Friedlander JS (2006) Management of retail buying. Wiley, Hoboken, NJ

Christopher M, Lowson R, Peck H (2004) Creating agile supply chains in the fashion industry. Int J Retail Distrib Manage 32(8):367–376

Cooper DR, Schindler PS (2014) Business and research methods. The McGraw-Hill Companies Inc, New York

Fernie J, Moore C (2003) Principles of retailing. Butterworth-Heinemann, Oxford

Fiorito SS, Gable M (2012) Retail buying: practices and policies in a global economy. Prentice-Hall, Upper Saddle River, NJ

Fiorito S, May E, Straughne K (1995) Quick response in retailing: components and implementation. Int J Retail Distrib 23(5):12–21

Fiorito SS, Gable M, Conseur A (2010) Technology: advancing retail buyer performance in the twenty-first century. Int Rev Retail Distrib Consum Res 38:879–893
Gabrielli V, Baghi I, Codeluppi V (2012) Consumption practices of fast fashion products: a consumer-based approach. J Fashion Mark Manage: Int J 17(2):206–224
Goworek H (2014) An investigation into retail buying roles and responsibilities for own-label clothing: a multiple case study. J Text Inst 105(7):760–769
Guercini S (2001) Relation between branding and growth of the firm in new quick fashion formulas: analysis of an Italian case. J Fashion Mark Manage 5(1):69–79
http://corporate.lcwaikiki.com/MainPage. Accessed Mar 2016
http://www.abet.org/. Accessed June 2016
http://www.iso.org/iso/home.htm. Accessed June 2016
http://www.ptc.com/product-lifecycle-management/windchill/flexplm. Accessed June 2016
http://www.wgsn.com/en/wgsn/. Accessed June 2016
http://www.polytropon.com/en/Software/MockShop-Virtual-Merchandising.html. Accessed June 2016
http://www.tgsd.org.tr/sektorel-bilgiler/raporlar/. Accessed Mar 2016
http://www.ihkib.org.tr/tr/bilgi-bankasi/raporlar/aylik-ihracat-bilgi-notlari/k-294. Accessed Mar 2016
Jacobsen ML (2009) The art of retail buying. Wiley, Singapore
Johansson U (2001) Retail buying: Process, information and IT use: a conceptual framework. Int Rev Retail Distrib Consum Res 11:329–357
Johansson J, Burt S (2004) Buying of private brands and manufacturer brands in grocery retailing: a comparative study of buying processes in the UK, Sweden and Italy. J Mark Manage 20:799–824
Levy M, Weitz BA, Beitelspacher LS (2012) Retailing management. McGraw-Hill/Irwin, New York
McGoldrick PJ (2002) Retail marketing, 2nd edn. McGraw Hill, Maidenhead
Quinlan C (2011) Business research methods. South-Western Cengage Learning, Andover
Rabolt NJ, Miler JK (2009) Concepts and cases in retail and merchandise management. Fairchild Books, New York
Schröder C (2015) The replication of retail fashion formats into foreign countries. Springer Gabler
Sheth JN (1973) A model of industrial buyer behaviour. J Mark 37:50–56
Sheth JN (1981) A theory of merchandise buying behaviour. In: Stamp RW, Hirschman EC (eds) Theory in retailing. American Marketing Association, pp 180–189
Sohal A, Perry M, Pratt T (1998) Developing partnerships and networks: learning for practices in Australia. Techovation 18(4):245–251
Swindley DG (1992) The role of the buyer in UK multiple retailing. Int J Retail Distrib Manage 20:3–15
Taplin IM (1999) Continuity and change in the US apparel industry: a statistical profile. J Fashion Mark Manage 3(4):360–368
Wall M, Sommers M, Wilcock A (1994) The retail buying of fashion goods: underlying themes of the sourcing process. Int Rev Retail Distrib Consum Res 4:177–193
Webster FE, Wind Y (1972) A general model for understanding organizational buying behaviour. J Marke 36:12–19
Zentes J, Morschett D, Schramm-Klein H (2011) Strategic retail management. Gabler Verlag Springer Fachmedien Wiesbaden GmbH

Part II
Methods, Technology and Fashion

A Conceptual Design of Intelligent Shoes for Pregnant Women

Eva Dimou, Athanasios Manavis, Evridiki Papachristou and Panagiotis Kyratsis

Abstract Design and technology are coming together to shape the future. Nowadays, researchers demonstrate that based on different technological fields, science is able to create a future that sometimes people find difficult to imagine. The present paper contributes towards the creation of new customized products, based on wearable technology, which is applied on mass consumption products. The case study used refers to the design of an intelligent, programmable shoe for pregnant women made of synthetic biological material (protocells). The paper's research methodology was based on a literature review and a series of interviews with some of the experts in the field of physiotherapy.

Keywords Intelligent shoes · Pregnant women · Product customisation · Wearable technology · Synthetic biological material

1 Introduction

Design and technology are coming together to shape the future. It is generally considered that the possibility of addressing a series of needs from both the psychological and emotional point of view, can lead to the production of wearable intelligent products.

Wearable manufacturers must focus on the product design as well as on the specifications and features of their products with an aim to impress their consumers.

E. Dimou
School of Art and Design, Coventry University, Coventry, UK

A. Manavis · P. Kyratsis (✉)
Department of Mechanical Engineering and Industrial Design,
Technological Education Institution of Western Macedonia, Kozani, Greece
e-mail: pkyratsis@teiwm.gr

E. Papachristou
School of Production Engineering and Management, Technical University of Crete, Chania, Greece

R. Rinaldi and R. Bandinelli (eds.), *Business Models and ICT Technologies for the Fashion Supply Chain*, Lecture Notes in Electrical Engineering 413,
DOI 10.1007/978-3-319-48511-9_6

Wearable technology needs further development in order to inspire fashion designers and create the products that users need. This could be achieved by slimming down the bulky elements of wearable technology (Kalinauckas 2015).

Technology in healthcare is advancing at a tremendous pace as designers continue to innovate with new materials, sensor technologies and engineering methods. Easy access to personalised healthcare information has also made the general public to take more control of managing their own health and demanding to do more of it (Daim et al. 2015).

2 Wearable Design and Technology

As wearable device is considered every device which is worn for an extended period of time, processes and controls its user's inputs and enhances his experience. A wearable can be categorized broadly into the following 6 categories:

- Lifestyle: includes smart watches, smart glasses and devices used for voice and video calling, gesture control etc.
- Entertainment: includes devices used for augmented reality, smart gloves, gesture, control devices etc.
- Medical: includes devices used for cardiac monitoring, hearing aid, bionics, remote monitoring of patients.
- Fitness: includes devices used for measuring heart rate, distance travelled, skin temperature.
- Gaming: includes devices that use augmented reality for gaming.
- Industrial: includes devices that help in hands-free and remote operation for business and industrial purposes.

Two years earlier, the trend was mostly towards activity monitoring. On the other hand, health has always been a driver of great importance for wearable technology. However, an enormous development is occurred today and consists of wearable products in the fields of fashion, hybrid functionality, smart glasses/watches, health and fitness (http://www.vandrico.com/wearables/ 2016).

Liu et al. designed, fabricated and tested a sweat-based conductivity sensor device towards real-time non-invasive physiological condition monitoring device for humans. All sweat collector, conductivity sensor and the interfacing circuit were developed to form a wearable device. Human testing was performed to prove the feasibility of the proposed sweat sensing system for the real-time non-invasive monitoring of human sweat (Liu et al. 2015).

Kim et al. developed a wearable chemical sensor based on a fingernail platform. Fingernails represent an attractive wearable platform because beauty products are merged with chemical sensing in order to enable monitoring of the surrounding environment. This novel idea can be expanded towards diverse analytes for various applications in connection to the design of the recognition layer (Kim et al. 2016).

Hunt and Hunt investigated the feasibility of a novel smartphone-based system for heart rate feedback control in outdoor running. An Android smartphone was employed, together with wearable wireless sensor for heart rate and running speed. The precision achieved lead to the conclusion that the system can be used for accurate achievement of prescribed exercise intensity for development and maintenance of cardiorespiratory fitness (Hunt and Hunt 2016).

Kwee-Meier et al. investigated innovations in ubiquitous and networked computing such as wearable locating systems for passengers that could enable faster muster and more efficient search for missing people. Passengers' acceptance of these technologies is crucial and as a result, a context-specific technology acceptance model was developed based on a literature review and qualitative interviews with passengers. The model was validated and presents insights into the relationships between passenger characteristics and context-specific factors (Kwee-Meier et al. 2016).

Son et al. explored computation-intensive real-time optical character recognition (OCR) and developed an adaptive power management scheme that predicts the execution time for OCR and minimises its energy consumption, while meeting its time constraint. During this research a popular open source OCR engine (Tesseract) was used to verify the proposed scheme (Son et al. 2016).

Zhang et al. developed a solution to power portable electronic devices in a wearable manner such as fabricating an all-solid photovoltaic textile. In the same way, that the plants absorb solar energy via photosynthesis, people can wear the as-fabricated textile in order to power small electronic devices (Zhang et al. 2016).

An et al. demonstrated a stretchable and transparent electrodes using CuZr metallic glasses in the form of nanotrough networks. A heater was presented and it was based on the metallic glasses nanotrough network with a wide operating temperature range and excellent stretchability. The proposed strategy suggests that next generation wearable electronic or automotive applications can be explored (An et al. 2016).

Kritzler et al. describe the concept and implementation of a safety system for personal protective equipment based on wearable sensors and wireless technology. The aim of the presented system is to ensure that the right personal protective equipment required for a specific task is worn. Results from interviews with users are also presented (Kritzler et al. 2015).

Lipovsky and Ferreira presented a rehabilitation strategy after a stroke. The strategy is supported by a novel robotized systems combined with virtual reality. Nowadays, these novel systems are expensive and still require human support. The proposed "self-rehabilitation system" is based on a robotic glove, an Arduino board, a Myo armband and a virtual reality video game made using Unity3D™. A low cost and easy to use hand rehabilitation system is achieved and offered for further studies (Lipovsky and Ferreira 2015).

The paper's research methodology was based on a series of interviews with experts in the field of physiotherapy. Additionally, the use of techniques such as deconstruction, manipulation, transformation, summarisation, generalization, abstraction, synthesis, benchmarking etc. was successfully applied. Based on the aforementioned methodology the design concept was developed. The main directions and characteristics of that innovative pair of shoes were the factors of adjustable size, negative heel technology and an easily fastened system to suitably solve the relevant issues. For the complete creation of a successful product, it was necessary to experiment with existing trends, forms, technologies and materials based on the philosophy of personalization. The idea of self-regenerating 3D-printed, synthetic biological soles with the use of protocells constitutes a new tendency for the creation of artificial materials by generating living organisms and it is based on real science.

3 Application to Pregnant Women

During pregnancy women's body undergoes huge changes hormonally and anatomically. Body weight and shape, and endocrine system are only some of the most common and obvious modifications. These changes may affect the balance and body stability. A series of activities in everyday life become more difficult and can cause discomfort and pain. Two are the most common problems that a pregnant woman is facing: swollen feet and back hip pain.

The predominant reason of this situation is the release of the hormone relaxin, which causes the ligaments of the foot to become lax and stretch out. That seems not to be the only reason. Both water retention and weight gain plays a dominant role as well. Furthermore, because of their bad posture, when the pregnancy progresses, most of the women are suffering from back, knee and hip pain. The pelvis tilts and the back arches help them keeping their balance. Poor posture occurs naturally from the stretching of the woman's abdominal muscles as the fetus grows, whereas these muscles are less able to contract and keep the lower back in proper alignment.

Wearing supportive shoes during pregnancy could be the key solution to the aforementioned issues of swollen feet and back, hip pain. Supportive shoes means to use shoes which support woman's arches without squeezing or pinching them. Sneakers are preferable instead of floppy ballet flats that many women are choosing. Additionally, compression socks or stockings are able to help this specific user group to control lower-extremity swelling, whereas supportive insoles are recommended by doctors or physical therapists and can be especially effective for users with low arches.

To ensure the validity of the proposed research, interviews with physiotherapists/experts were conducted. The nature of the questions made aimed in targeting and determining the basic demands and thus conclude on the necessary design aspects. Questions were carried out about the different lifestyles and habits according to the gait and how these could affect people's health or wellbeing, as well as if people attach sufficient importance when buying footwear. Also, questions were raised about the main functions that a shoe has to fulfill in order to avoid any user's injury, especially with respect to pregnant women.

According to the answers collected, the concept of the design was identified. Some points, on which particular attention is not given traditionally, were highlighted and constituted main inspiration for the incorporation of new properties and characteristics to the final product. So, despite of the meaning of wearing supportive and comfortable shoes, preferably orthopedic, the importance of the right insoles was emphasized. Additionally, specifically for pregnant women, that's more significant because of circulation problems which most of them are facing. Nevertheless, shoes have to be able to accommodate their feet by being light, practical, and easy to put on and having the right size. Flexibility is another specification that should be included, so that the shoes can afford the result of the relaxed ligaments, which are caused because of the released hormones in the blood during the pregnancy period. Furthermore, the reference to a technology called negative heel contributed significantly to the redesign of the product. Three additional important factors identified were the (a) importance of foot-muscle exercises or massages, (b) parameter of hydration and (c) easiness of shoes fastening mainly because of the changes in pregnant women body.

4 Conceptual Design

Based on the above research, the aim was to design a customized, supportive pair of shoes for pregnant women. The product is expected to follow the philosophy of both wearable design technology and mass customization. For the successful fulfillment of the aforementioned objective, it was absolutely necessary to experiment with existed trends, forms and geometries of corresponding products. According to the results of the literature research and interviews, the requirements of this particular user group from such a product are very specific and when narrowing down the basic needs of this case study and their solutions, three directions are considered to be crucial.

The factor of shoe's adjustable size: based on the appearance of swollen foots and circulation problems, it is important and helpful for pregnant women to wear a flexible, comfortable and evolving shoe. These characteristics were incorporated in a series of conceptual designs presented in Fig. 1 (i.e. futuristic geometry, adjustable fixed mold shoe, back lacing solution, flexible, washable slip-on, simple and modern design).

The negative heel technology: by the means of this technology, it is possible for pregnant women to stand straight naturally. They can easily lift their toes a little higher and push their body backwards. As a result, back and hip pain can noticeably be reduced (Fig. 2).

The easiness of shoes fastening: it is difficult for these particular users to successfully set their shoes, especially at the last months of pregnancy. That is the main reason why emerged the need of an easy to slip shoe and get tied (Fig. 3).

The conceptual design presented in Fig. 4 is a creative combination of the three aforementioned ideas. The aspects of functionality, new technologies, perfect

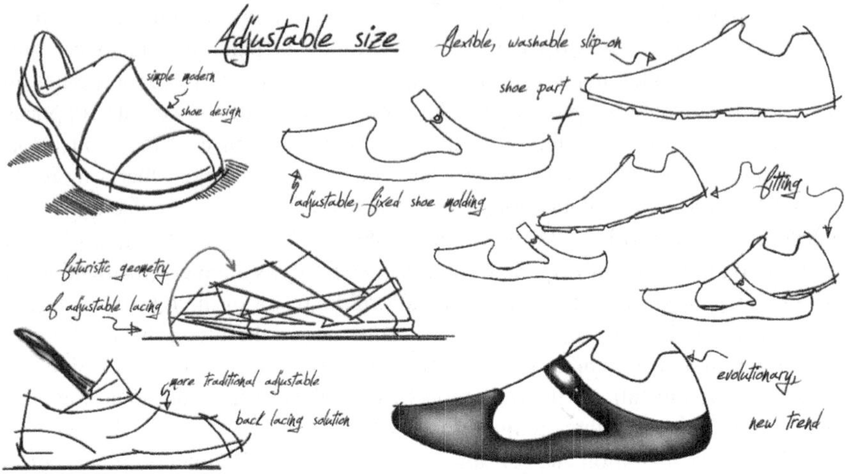

Fig. 1 Conceptual design based on the demand for adjustable sized shoes

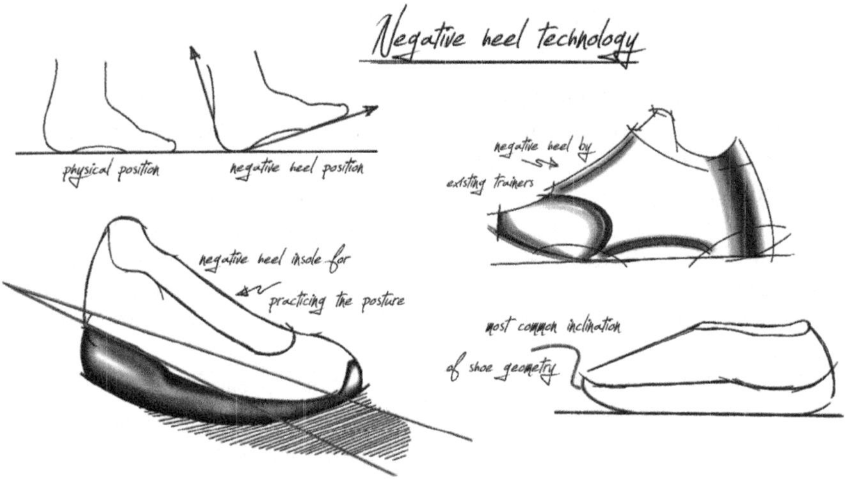

Fig. 2 Conceptual design based on negative heel technology

fitting, material selection, form and efficiency were included. Round slipper geometry with relative high insoles were incorporated. Following the rules of negative heel technology, by embracing the foot and slipping easily into it, a new shoe design was created. The form's curves bring out the product's feminine nature, since it is strictly for women designed. At the same time, they create a recess at the back side of the shoe. Thus the user can take out the shoe, simply by exerting some

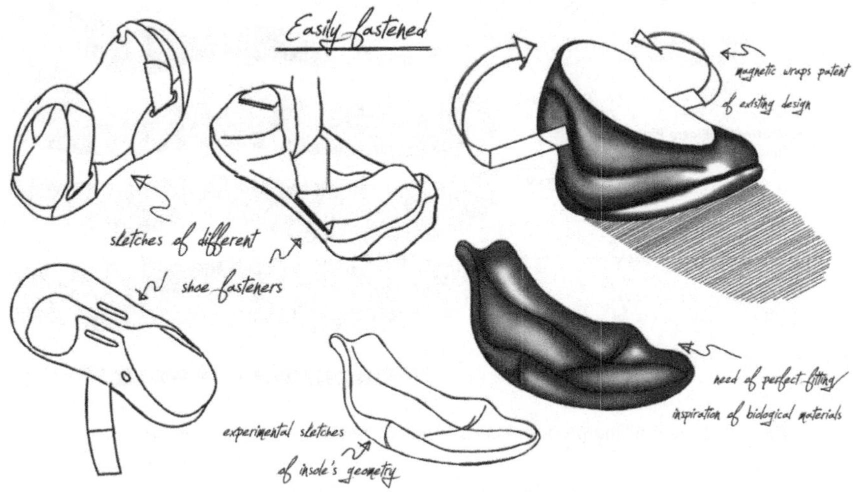

Fig. 3 Conceptual design based on the easiness of shoe fastening

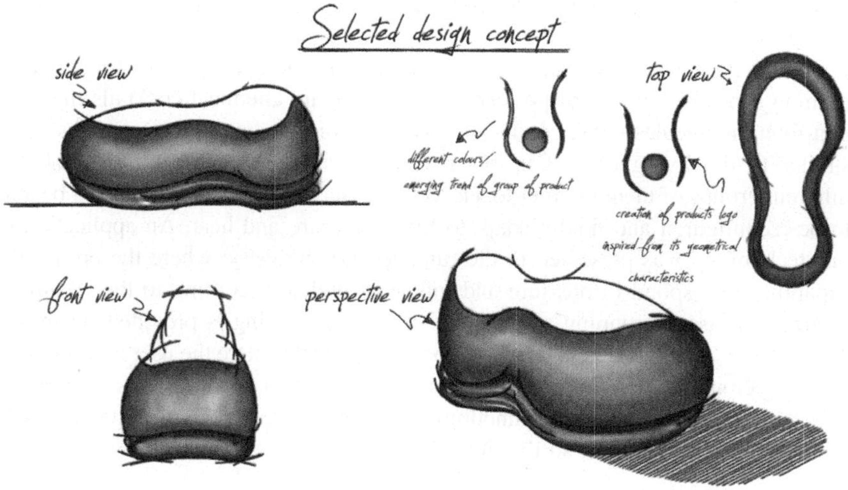

Fig. 4 Conceptual design for pregnant women

force (with the help of the other foot). For the material selection a 3D printed flexible polymer called "Elasto Plastic" is proposed. Due to this polymer, the perfect fit and customization of individual sizes will be possible after the appropriate 3D scanning process, which is already a common practice nowadays (http://www.shapeways.com/blog/archives/2084-introducing-maker-materials-the-new-improved-elasto-plastic.html 2016).

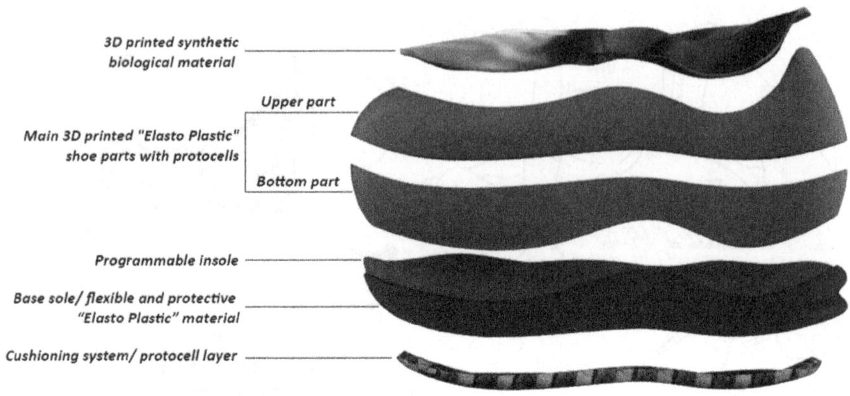

Fig. 5 Exploded view of the proposed conceptual design and its functionalities

Foot-muscle exercises and massages as well as the right hydration and water pressure on the skin are able to reduce or even lead to the disappearance of the swollen feet phenomenon. The main idea is to design a programmable insole, which will be able to massage user's feet and at the same time to hydrate them and create the appropriate temperature.

Researchers have proposed the creation of self-regenerating soles for a pair of running shoes by the use of protocells. Protocells are chemical cocktails made by non-living molecules in lab conditions. By combining them, they create substances similar to the living cells. For this reason, scientists are now mixing together different groups of them in order to succeed in behaving like living cells and be able to be reconfigured and finally adapt to light, pressure and heat. An application of this technology was presented in the running shoe industry; where the protocells' capability to respond to pressure and inflate or deflate according to the texture of ground, the user is running while more or less cushioning is provided (Dent and Sherr 2014). The vision is to create a product's system, where the user will not have an indirect interaction (via smart phone or computer) in order this to be customizable. A system able to be functioning only with user's individual touch, smell, voice, heart bit, cells etc., so that it will be absolutely personalized (Fig. 5).

5 Conclusions

Wearable technology is already a trend in product design. Customized products with increased added value are on high demand. With the rapid expansion of technological advancements more opportunities are given to fashion designers. Combining fashion design principles with a series of technologies, that become available, users can acquire personalized products in lower cost.

A case study concerning the pregnant women shoe design needs is presented. All conceptual designs contribute towards the successful solution of both the main problems they face, i.e. swollen feet and back hip pain. The combination of design principles, material sciences and engineering methodologies offer a new perspective in the area. New customized supportive shoes can be offered to pregnant women and relief their pains.

References

An BW, Gwak EJ, Kim K, Kim YC, Jang J, Kim JY, Park JU (2016) Stretchable, transparent electrodes as wearable heaters using nanotrough networks of metallic glasses with superior mechanical properties and thermal stability. Nano Lett 16:471–478

Daim TU, Al-Mulla N, Sengupta SB, Shah K, Demchig B (2015) Technology assessment: case of the wearable computing for fitness. Int J Med Eng Inform 7(4):321–364

Dent AH, Sherr L (2014) Material Innovation: product design. Thames and Hudson, London

http://www.shapeways.com/blog/archives/2084-introducing-maker-materials-the-new-improved-elasto-plastic.html. Accessed 1 March 2016

http://www.vandrico.com/wearables/. Accessed 1 March 2016

Hunt KJ, Hunt AJR (2016) Feedback control of heart rate during outdoor running: a smartphone implementation. Biomed Signal Process Control 26:90–97

Kalinauckas AT (2015) Wearable technology. Eng Technol 10:36–43

Kim J, Cho TN, Valdes-Ramirez G, Wang J (2016) A wearable fingernail chemical sensing platform: PH sensing at your fingertips. Talanta 150:622–628

Kritzler M, Backman M, Tenfalt A, Michahelles F (2015) Wearable technology as a solution for workplace safety. In: 14th international conference on mobile and ubiquitous multimedia (MUM 2015), Centerlinz, Austria

Kwee-Meier ST, Butsler JE, Schlick C (2016) Development and validation of a technology acceptance model for safety-enhancing, wearable locating systems. Behav Inf Technol. doi:10.1080/0144929X.2016.1141986

Lipovsky R, Ferreira HA (2015) Self hand-rehabilitation system based on wearable technology. In: 3rd workshop on ICTs for improving patients rehabilitation research techniques (REHAB 2015), Portugal

Liu G, Ho C, Slappey N, Zhou Z, Snelgrove SE, Brown M, Grabinski A, Guo X, Chen Y, Miller K, Edwards J, Kaya T (2015) A wearable conductivity sensor for wireless real-time sweat monitoring. Sens Actuat B Chem 227:35–42

Son S, So H, Kim J, Choi D, Lee HJ (2016) Energy-efficient adaptive optical character recognition for wearable devices. Electron Lett 52:113–115

Zhang N, Chen J, Huang Y, Guo W, Yang J, Du J, Fan X, Tao C (2016) A wearable all-solid photovoltaic textile. Adv Mater 28:263–269

References

The Adaptive Fitting Room

Børge Sjøbakk, Andreas Dypvik Landmark and Hans Petter Hübert

Abstract In order to remain attractive and economically viable whilst supplementing online presence, physical stores need to play to their strengths and engage customers in novel ways. Stores have the possibility to act as experimental arenas, where customers can browse apparel through interactive installations and try on products in semi-realistic situations. For fashion retailers, however, the average fitting room experience is rarely representative of actual use. The product range is often diverse and varies with seasons, which makes it difficult to create a one-experience-fits-all fitting room. In this paper, we present a concept for a new customer experience. By exploiting automatic single item identification using RFID, we propose a fitting room that adapts to the products that enter the fitting room, using video projection mapping, magic mirrors and screens, lighting and sound. This allows different types of customer engagement prior to the point of purchase, opening up new possibilities for integrating product information, recommendations, clienteling and social media. Some expected benefits are illustrated through a set of user scenarios.

Keywords Fitting room · RFID · Retail · Experience · Physical store

1 Introduction

Today's fashion retailers face multiple challenges, most of which are related to the expansion into and balancing of omni-channel retailing. Customers are more informed than before, and the transaction cost of comparing and switching between both online and brick and mortar retailers is decreasing (Nunes and Cespedes 2003;

B. Sjøbakk (✉) · A.D. Landmark
SINTEF Technology and Society, Department of Industrial Management,
Trondheim, Norway
e-mail: borge.sjobakk@sintef.no

H.P. Hübert
Moods of Norway, Oslo, Norway

R. Rinaldi and R. Bandinelli (eds.), *Business Models and ICT Technologies for the Fashion Supply Chain*, Lecture Notes in Electrical Engineering 413,
DOI 10.1007/978-3-319-48511-9_7

Stone et al. 2002). Physical stores have traditionally played the advantage of enabling product 'touch and feel', correct fit and sizing, zero delivery lead time, and advice and up-sales from sales representatives (Bhatnagar et al. 2000; Liao and Cheung 2001; Levin et al. 2003). However, the lines between online and offline retail are fading (Enders and Jelassi 2000). Online channels increasingly offer free returns, next or same day shipping and improved choice navigation aiding the customers. At the same time, online retail benefits from advantages such as consolidated inventory, economies of scale in purchasing and lower labor and facilities costs.

In order to remain attractive and economically viable whilst supplementing online presence, physical stores need to play to their strengths and engage customers in novel ways (Baker et al. 2002). In this respect, physical stores have a possibility to act as experimental arenas, where customers can browse apparel through interactive installations and try on products in semi-realistic situations. The latter is perhaps more predominant within sporting goods and recreational equipment. For example, American REI offer customers in their Seattle flagship store the opportunity to test climbing equipment in a 20 m instore climbing wall (REI 2016). For fashion retailers however, the average fitting room experience is rarely representative of actual use. The product range is often diverse and varies with seasons, which makes it difficult to create a one-experience-fits-all fitting room; for instance, a cocktail dress is obviously used in other environments than a heavy winter coat, but they are still often found in the same store within a season. Further, they are usually tried on under the same ambience in shared fitting rooms.

In this paper, we present a concept for a new customer experience—*the adaptive fitting room*. By exploiting automatic single item identification using RFID, we propose a fitting room that adapts to the products that enter the fitting room, using video projection mapping, magic mirrors and screens, lighting and sound. This allows different types of customer engagement prior to the point of purchase, opening up new possibilities for integrating product information, recommendations, clienteling and social media. This will possibly attract customers to physical stores and augment the stores' online presence. Further, we present a discussion in which we motivate such a concept through showing how such an investment makes sense and affects upstream actors of the retail value chain, too.

Several authors (e.g. Loebbecke et al. 2008; Uhrich et al. 2008; Choi et al. 2015; Melià-Seguí et al. 2013; Serra et al. 2011) describe the use of RFID to provide recommendations and information about garments inside the fitting room. However, to the best of our knowledge the use of such technology to trigger an alteration of the fitting room's ambience has not previously been proposed. Some expected benefits of the adaptive fitting room are illustrated through a set of user scenarios. In the process of developing these scenarios, interviews with the retail manager, the head designer and a shop floor manager of a Norwegian fashion retail chain have been carried out.

The remainder of the paper is structured as follows. First, the adaptive fitting room is described. Here, a brief summary of previous research on fitting rooms and interactive installations is provided, before outlining the envisioned functionality of the concept. A set of user scenarios is provided to illustrate its use and further motivate following discussion. Thereafter, identified challenges and future prospects for research are discussed.

2 The Adaptive Fitting Room

2.1 Theoretical Background

The shop environment's effect on customer behavior has received substantial attention by researchers and practitioners (see e.g. Bäckström and Johansson 2006). Some suggest that only a limited number of consumer choices are based on conscious information processing strategies; the rest is said to be caused by unconscious effects of all kinds of cues in the environment (Dijksterhuis et al. 2005). While many atmospheric variables can affect the shopping experience, lighting has been the focus of many studies, as it can easily be altered to create different moods (Baumstarck and Park 2010). However, as people perceive the world through all their senses simultaneously, other sensory stimuli such as touch, sound and smell can also influence environments and improve the shopping experience and behavior (Soars 2009). In the same way that atmospherics in the main store can affect the shopping experience either positively or negatively, the dressing room atmosphere can be crucial to making a purchase (Baumstarck and Park 2010). Therefore, we argue that an alteration of the fitting room ambience is highly relevant in physical stores' quest for better customer experiences.

When looking beyond 'traditional' fitting room qualities, such as spaciousness and lighting (Baumstarck and Park 2010), most of the recent research on fitting rooms can be divided into three main streams: (1) interactive installations, (2) virtual try-on and, (3) RFID in fitting rooms. The first stream, *interactive installations*, covers aspects such as conceptualizations of, and challenges with, multimedia mirror systems for physical stores (Begole et al. 2009; Zhang et al. 2010); more overall concerns regarding risks and benefits of interactive installations (Campos et al. 2011; Akpan et al. 2013); and, other fitting room concepts such as *the socially-interactive dressing room,* which integrates social media in the fitting room (Liew et al. 2011). The second stream, *virtual try-on*, looks beyond the physical garment and tries to develop virtual fitting room solutions for use either in-store or online. This stream has a strong technological focus, with many authors focusing on image processing and augmented reality technologies (Chang et al. 2013; Kjærside et al. 2005; Traumann et al. 2015). The third stream, *RFID in fitting rooms*, takes into consideration the progression of item-level RFID and looks at how this opens for detailed monitoring of visual merchandising efficiency, correction of individually misplaced items and instore product flow (Loebbecke et al. 2008; Choi et al. 2015) and enables interaction with e.g. smart dressing rooms, displays and mirrors to improve the shopping experience (Loebbecke et al. 2008; Uhrich et al. 2008; Choi et al. 2015; Melià-Seguí et al. 2013; Serra et al. 2011).

2.2 The Fitting Room Concept

The main idea behind the adaptive fitting room is to use the RFID tag of each garment to identify which garments the customer is bringing into the fitting room. Based on identification of the product, the customer gets a matching experience by automatically altering the fitting room's ambience. The fitting room should mimic environments in which the products are typically used, but could also be altered in response to user inputs. This is illustrated in Figs. 1, 2 and 3.

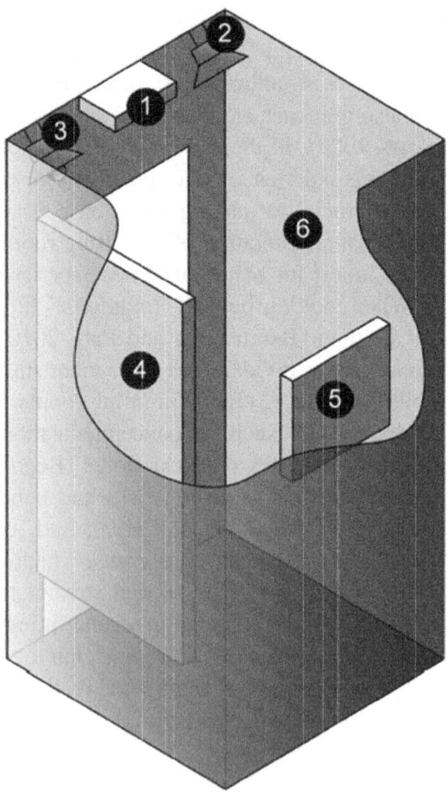

Fig. 1 The adaptive fitting room when it is not in use. Foreseen functionality includes (*1*) RFID antenna; (*2*) speaker; (*3*) lighting; (*4*) intelligent (magic) mirror; (*5*) information/omni channel panel; (*6*) (projection) screen

Fig. 2 The adaptive fitting room when in use, showcasing the change of ambience given when a customer brings a certain garment into the fitting room

Fig. 3 The customer sees itself as standing in a landscape when trying on clothes for skiing. Product information can be provided through the magic mirror

2.3 User Scenarios

In this section, we describe possibilities that an adaptive fitting room might represent for four key roles in the retail supply chain; the Customer, the Shop floor assistant, the Retail manager, and the (product) Designer.

2.3.1 Customer

The modern customer is making a conscious choice to go to a brick-and-mortar store over other retail channels. While the reasons vary, the store of the future will certainly play to its strengths by attracting customers based on the advantages of brick-and-mortar and high-fidelity customer engagement limited to the physical realm.

The customer browses the store and selects a few items that he brings to the adaptive fitting room. The fitting room immediately detects the items the customer has brought into the room through RFID. The customer is looking to purchase a winter coat, but has not made a final decision. The fitting room recognizes the product category and adapts both the lighting and scenery to place the customer in a bright and wintery landscape—both in order to augment the shopping experience, but also to place the product in a scenario that is realistic to the actual end-use of the product.

Undecided as he is the customer uses the in-room system for choice navigation—looking at the alternatives sizes, colors and products the store can offer. The shop floor assistant is on hand and supplying the alternative size. The in-room system also offers recommendations based on customer club membership and previous purchases the customer has made.

After finding a product, through the aid of the in-room navigation and shop floor assistant, with a desired fit and color, the customer opts for home delivery rather than carrying the bulky winter coat home himself. This order is actually filled from a different store that has a larger inventory and better facilities for home delivery, but this is invisible for the user who receives it at home at the agreed upon timeslot.

2.3.2 Shop Floor Assistant

The shop floor assistant may monitor real-time flow of products in the store based on the RFID-equipment fitted in the store. This allows for both an up-to-date inventory as well as the ability to see which products are tried in the various fitting rooms. For customers who have chosen to allow self-identification, it is also possible to for the assistant to see what the customer already owns from the brand (regardless of which channel it was obtained through).

Upon seeing that a customer has entered the adaptive fitting room with a winter coat, the assistant may choose to use the electronic recommendation engine to calculate potential items that he can recommend to the customer as well as alternate fits if he believes the customer might want to try on a different size. This reduces the "lead time" for the assistant in offering alternative or additional sizes or products to the customer. Additionally, the electronic aid helps the shop floor assistant in classifying how to approach the customer for a best possible interaction, as well as prioritizing between different customers based on more data.

2.3.3 Retail Manager

RFID already allows the retail manager to monitor the inventory stock levels and gauge the efficacy of the visual merchandising and layout of the store. The detailed flow of products also better track to which extent products cannibalize each other—allowing the manager to differentiate between products "swapped on the rack" or if they make it to the fitting room (and later if it converts to a sale).

The adaptive fitting room not only gives an indication of fitting-room versus non-fitting room conversion rates, but also allows the manager to go "beyond" Point-of-Sale for conversion rates and distinguish between "which products are tried, but never sold" and "which products never make it off the shelf". The aggregation of this customer behavior allows for more detailed customer profiles and constant profiling and adapting the content of the fitting rooms to create the illusion of "a spit wash for every customer" based on detailed profiles.

The low cost of changing the fitting room experience when deployed also allows to continually re-create the initial wow-factor and deliver customer experiences that attracts potential customers into the store.

2.3.4 Designer

In general, the designer is less 'in-the-loop' than the other roles described, often working a season or two ahead of the products that are currently in store. Feedback for designers often comes through multiple sources such as peer- and expert reviews, as well as turnover statistics of which products have sold well.

Based on fitting room statistics, such as combinations and products tried, but not bought—the adaptive fitting room may open up for pinpointed surveys of customers trying on specific products or combinations—for direct customer engagement.

3 Discussion and Prospects for Future Research

The adaptive fitting room is an ambitious concept which is founded on numerous assumptions. First, it assumes that customers are interested in trying on garments in an environment that simulates situations in which the garment would typically be worn if purchased. Second, it assumes that an alteration of the fitting room is considered to be a positive experience for the customers. Third, it assumes that a fitting room provides sufficient space for numerous sensory stimuli. Fourth, it assumes that RFID can be used to trigger the adaption of the fitting room. Fifth, it assumes that the adaptive fitting room pays off. The list goes on. These assumptions need to be taken into consideration when adopting the concept.

Some of the assumptions relate to customers' willingness to adopt the concept. After all, more and more retailers use advanced techniques in order to create compelling in-store experiences to their customers, while the same customers regard traditional values such as the sales personnel's behavior, satisfactory product ranges and a good store layout as determinants of their in-store experience (Bäckström and Johansson 2006). We believe that the adaptive fitting room is not for everyone. It is aimed at companies that (should) try something 'crazy' once in a while, which have customers that crave technology and innovation rather than what is familiar. If a retailer that does not fit this profile adopts the adaptive fitting room, we fear that it may become just another 'irritating aspect of the shopping environment' (see d'Astous 2000).

Other assumptions relate to technical and physical feasibility of the concept. In our research, RFID antennae have been installed in real-life fitting rooms of a retailer. We see that detecting garments in one room at the time, with customers blocking the tags from the antennae when trying on garments, is a real challenge. This is important to overcome, as altering the fitting room to a real-life simulation requires a steady read over some time to ensure that the customer is in fact inside the fitting room. Other challenges are how to contain sound and light within fitting rooms that need may need to be open at the top and bottom due to theft protection and safety, and how to create a realistic environment, either with screens or video projection, within the very limited space of a typical fitting room.

Finally, it is a question of payback. While some (e.g. Soars 2009) argue that enhanced shopping experiences can have a significant impact on decision-making, store choice and spend, it is hard to foresee the effects of the adaptive fitting room with respect to actual conversion and increased sales. Even when in place, the effects of the concept may be hard to quantify, as it may for example bring customers to the store that purchase something without trying it on.

As is evident, there are numerous prospects for future research related to the adaptive fitting room. The multitude of assumptions needs to be addressed in order to further develop the concept.

4 Conclusion

In this paper we have presented a concept for a new customer experience—*the adaptive fitting room*. The idea behind this concept is to exploit automatic single item identification using RFID to adapt the fitting room of a retail store to the products that enter it, using video projection mapping, magic mirrors and screens, lighting and sound. In addition to opening up new possibilities for integrating product information, recommendations, clienteling and social media in the fitting room, we believe that it will possibly attract customers to physical stores and augment the stores' online presence. This is important in a time where physical stores need to play to their strengths and engage customers in novel ways in order to remain attractive and economically viable.

The adaptive fitting room concept acknowledges the potential of multiple sensory stimuli to influence shoppers' behavior. Further, it assumes that people are engaged by interactive installations, and that virtual simulations cannot fully replace physical touch and feel when it comes to trying on garments. Finally, it is strongly based on single-item RFID being in place. As such, it augments two of the three identified research streams on fitting rooms.

Acknowledgements This research has been conducted within the project Reflect your Moods funded by the Research Council of Norway. The authors would like to thank the participants of the project for providing valuable empirical data. We would also like to thank Geir Vevle of HRAFN for useful discussions on technical solutions.

References

Akpan I, Marshall P, Bird J, Harrison D Exploring the effects of space and place on engagement with an interactive installation. In: Proceedings of the SIGCHI conference on human factors in computing systems, 2013. ACM, pp 2213–2222

Baker J, Parasuraman A, Grewal D, Voss GB (2002) The influence of multiple store environment cues on perceived merchandise value and patronage intentions. J Mark 66(2):120–141

Baumstarck A, Park NK (2010) The effects of dressing room lighting on consumers' perceptions of self and environment. J Inter Des 35(2):37–49

Begole B, Matsumoto T, Zhang W, Yee N, Liu J, Chu M (2009) Designed to fit: challenges of interaction design for clothes fitting room technologies. In: Human-computer interaction. Interacting in various application domains. Springer, pp 448–457

Bhatnagar A, Misra S, Rao HR (2000) On risk, convenience, and Internet shopping behavior. Commun ACM 43(11):98–105

Bäckström K, Johansson U (2006) Creating and consuming experiences in retail store environments: comparing retailer and consumer perspectives. J Retail Consum Serv 13(6): 417–430

Campos P, Campos M, Jorge J (2011) How high can expectations go?: practitioner issues and risks of interactive installations. Interactions 18(3):30–35

Chang H-T, Li Y-W, Chen H-T, Feng S-Y, Chien T-T (2013) A dynamic fitting room based on microsoft kinect and augmented reality technologies. In: Human-computer interaction. Interaction modalities and techniques. Springer, pp 177–185

Choi S, Yang Y, Yang B, Cheung H (2015) Item-level RFID for enhancement of customer shopping experience in apparel retail. Comput Ind 71:10–23

d'Astous A (2000) Irritating aspects of the shopping environment. J Bus Res 49(2):149–156

Dijksterhuis A, Smith PK, Van Baaren RB, Wigboldus DH (2005) The unconscious consumer: effects of environment on consumer behavior. J Consum Psychol 15(3):193–202

Enders A, Jelassi T (2000) The converging business models of Internet and bricks-and-mortar retailers. Eur Manage J 18(5):542–550

Kjærside K, Kortbek KJ, Hedegaard H, Grønbæk K (2005) ARDressCode: augmented dressing room with tag-based motion tracking and real-time clothes simulation. In: Proceedings of the central european multimedia and virtual reality conference

Levin AM, Levin IR, Heath CE (2003) Product category dependent consumer preferences for online and offline shopping features and their influence on multi-channel retail alliances. J Electron Commer Res 4(3):85–93

Liao Z, Cheung MT (2001) Internet-based e-shopping and consumer attitudes: an empirical study. Inf Manage 38(5):299–306

Liew JSY, Kaziunas E, Liu J, Zhuo S Socially-interactive dressing room: an iterative evaluation on interface design. In: CHI'11 extended abstracts on human factors in computing systems, 2011. ACM, pp 2023–2028

Loebbecke C, Huyskens C, Gogan J (2008) Emerging technologies in the service sector: an early exploration of item-level RFID on the fashion sales floor. In: Information technology in the service economy: challenges and possibilities for the 21st century. Springer, pp 189–198

Melià-Seguí J, Pous R, Carreras A, Morenza-Cinos M, Parada R, Liaghat Z, De Porrata-Doria R (2013) Enhancing the shopping experience through RFID in an actual retail store. In: Proceedings of the 2013 ACM conference on pervasive and ubiquitous computing adjunct publication, 2013. ACM, pp 1029–1036

Nunes PF, Cespedes FV (2003) The customer has escaped. Harvard Bus Rev 81(11):96–105

REI (2016) Climbing the pinnacle at seattle REI. https://www.rei.com/stores/seattle/climb-class.html. Accessed 4 March 2016

Serra CC, Medeiros CR, Costa JR, Fernandes C (2011) Mirror-integrated transparent antenna for RFID application. Antennas Wirel Propag Lett IEEE 10:776–779

Soars B (2009) Driving sales through shoppers' sense of sound, sight, smell and touch. Int J Retail Distrib Manage 37(3):286–298

Stone M, Hobbs M, Khaleeli M (2002) Multichannel customer management: the benefits and challenges. J Database Mark 10(1):39–52

Traumann A, Anbarjafari G, Escalera S (2015) A new retexturing method for virtual fitting room using kinect 2 camera. In: Proceedings of the IEEE conference on computer vision and pattern recognition workshops, 2015. pp 75–79

Uhrich F, Sandner U, Resatsch F, Leimeister JM, Krcmar H (2008) RFID in retailing and customer relationship management. Commun Assoc Inf Syst 23(1):13

Zhang W, Begole B, Chu M (2010) Asynchronous reflections: theory and practice in the design of multimedia mirror systems. Multimedia Syst 16(4–5):293–307

Towards Case-Based Morphological Classification for Fashion Product Development

Thomas Fischer, Konrad Pfleiderer, Alexander Artschwager, Anke Rissiek, Magdalena Mandalka, Andreas Seidl and Rainer Trieb

Abstract This paper describes a novel approach for the morphological classification based on body scan data. This is an important topic for the fashion industry because traditional sizing systems do not provide perfectly fitting clothes. That is due to the diversity of human bodies and the limited number of primary dimensions in sizing. The idea is to have experts classifying a relevant number of representative body scans and to use this as a case and knowledge base for further automatic classification. This Case-Based Reasoning (CBR) approach is suitable especially for classification problems where many variables are involved and no clear rules can be identified. A new scan is compared with the already classified scans in the case base. The most similar ones are retrieved and their classification adapted to classify the new scan. A sophisticated similarity model ensures that only useful, relevant cases are retrieved from the case base. The system has been tested, including statistical analysis and field tests. An industry project confirmed the

T. Fischer (✉) · K. Pfleiderer · A. Artschwager
Centre for Management Research, German Institutes of Textile and Fiber Research, Denkendorf Körschtalstr. 26, 73770 Denkendorf, Germany
e-mail: thomasvfischer@ditf-mr-denkendorf.de

K. Pfleiderer
e-mail: konrad.pfleiderer@ditf-mr-denkendorf.de

A. Artschwager
e-mail: alexander.artschwager@ditf-mr-denkendorf.de

A. Rissiek · M. Mandalka · A. Seidl · R. Trieb
Human Solutions GmbH, Europaallee 10, 67657 Kaiserslautern, Germany
e-mail: anke.rissiek@human-solutions.com

M. Mandalka
e-mail: magdalena.mandalka@human-solutions.com

A. Seidl
e-mail: andreas.seidl@human-solutions.com

R. Trieb
e-mail: rainer.trieb@human-solutions.com

R. Rinaldi and R. Bandinelli (eds.), *Business Models and ICT Technologies for the Fashion Supply Chain*, Lecture Notes in Electrical Engineering 413,
DOI 10.1007/978-3-319-48511-9_8

feasibility of the approach and now helps the fashion company to adjust their product development to the real needs of the customers. The approach will be used in body measurement portals to optimize size tables and thus the fitting of the clothes as well as in e-shops to recommend the morphologically best garments to a customer.

Keywords Morphological classification • Case-based reasoning (CBR) • Bodyscanning • Morphotypes • Size table development • Fit optimization • Size recommendation

1 Introduction and Outline

As human populations change constantly up-to-date size surveys are the precondition for fitting fashion products and a high market share in the apparel industry. Numerous survey results from different countries and special target populations are used to develop standard sizing systems as well as to optimize grading and size ranges. Today survey results are provided in body measurement portals allowing standard statistical analysis as well as company-specific size chart development for key markets (Webster et al. 2012; Rissiek and Trieb 2012). Due to economic reasons the apparel industry tries to maximize their global market potential with a relatively small number of sizes. The standard sizes are often just defined by a few primary dimensions although body shapes, proportions and measurement correlations differ. Body scans from surveys show the variation of body shapes within one size and are therefore an optimal starting point for the definition and differentiation of morphotypes and at the same time for a better fit (Fig. 1).

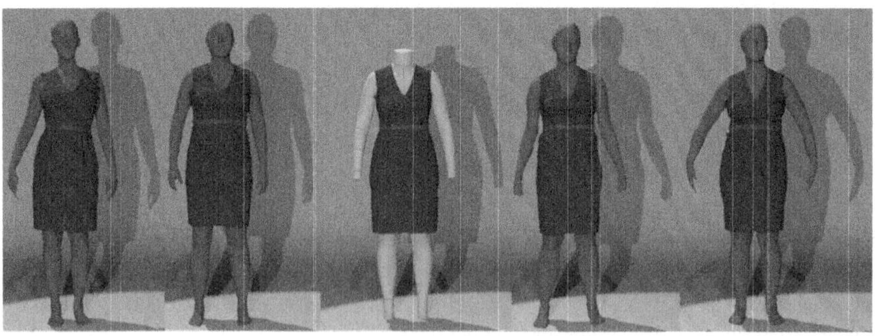

Fig. 1 Women in German size 42 with different body shapes in comparison to standard fashion manikin in size 42

The objective of the project is the morphological classification based on body scan data to be used for size system development and to provide better fitting clothes. The morphotype classification can be applied in online retail shops for size recommendation and curated shopping. Furthermore future collection development can incorporate the knowledge about morphotypes and provide special collections that optimally fit to individual morphotypes.

The analysis of the available body scan data showed that it is merely impossible to derive rules for the automatic morphological classification. On the other hand, experts are able to classify shapes when looking at different scans (front view, side view etc.). The Case-Based Reasoning (CBR) approach seems therefore rather promising. The idea is that experts evaluate visually a representative number of selected data sets (body scans) according to the newly developed morphological classification scheme. Automated, case-based classification then takes advantage of the expert knowledge by comparing a new scan with already classified ones and by deriving the new classification from the most similar ones.

The paper therefore reflects this approach and is organized as follows: The second chapter after this introduction analyses current morphological classification systems and their deficits. A new scheme is derived and described. Next, the chapter on Case-Based Reasoning for morphological classification goes into the details of how to apply CBR for the classification problem at hand. The next chapter presents some first results from experiments with larger numbers of cases and describes the benefits for garment making industry and for the end users. A brief summary concludes this paper.

2 Morphological Classification

Over the past years there have already been several approaches to characterize morphotypes (without using CBR methodologies) and to develop a classification scheme with respect to the needs of the Fashion industry which have some significant deficits: Most of them are only regarding the full body, and are therefore not appropriate for the classification of single body parts as upper or lower body that are indispensable for fit optimization (Bastos and Sabrá Senai Cetiqt 2014; Morlock et al. 2009). Furthermore they classify only by analyzing length and width dimensions and not regarding the complete 3D shape from body scanners (Duffy 1987; Surville et al. 2010). Their classification schemes have only been developed by regarding subjective and visual assessments. Most of them consider only single measurements, but no measurement relations and correlations between different classifications are accounted. As they altogether relied on expert classification which comes along with an enormous expenditure of time they are not validated on a higher number of test persons (Webster et al. 2012).

Morphological classification on base of CBR technology makes great demands on the configuration of the morphological classification system: The case base for expert classification has to include numerous body shapes to be appropriate for global markets as well as for specific target groups. The system needs to be scalable so that each of the classification levels can be used independently either for full bodies or single body parts as well as for all fashion product categories with their different requirements on regarded measurements and shapes. A simple extensibility is essential for transferability on up-to-date data pool coming from survey with body-scanning in different countries and on diverse ethnic groups. The consideration of the 3-dimensional human body needs provision for length, width, girth, depth measurements and body ankles. The additional regard to visual aspects of the 3D-shapes cannot be simply derived from body measurements but needs to regard various correlations and appropriate proportions. The morphological classification system has to be validated on a statistically representative data base. Various approaches have proven that a maximum of three different characteristics within one classification level can be differentiated by the experts to ensure unique expert classifications.

The starting point for the development of the classification scheme was a detailed analysis of the variance of different key measurements in correlation with the most important primary dimensions. If the variance is not covered by the actual range between the sizes an additional morphotype classification is needed. Based on literature and scan review, a morphological classification scheme was developed. It comprises the following 10 features and is according ordinal scales.

Classification Full Body: Basic shape, relation shoulder and hip, Waist shape, Posture, Waist position
Classification Upper Body: Arm shape, Shoulder slope, Relation bust and back
Classification Lower Body: Leg Length, Leg shape, Buttock shape

Thorough analysis revealed that an ordinal, categorical scale fits best. This means that 3 (for one feature only 2) possible values are defined and the values have a natural order. This is an important point because otherwise operations like variance and mean value would not be feasible.

A web-based application allows remote classification for invited experts. They all had to select the most suitable value without seeing others' results, no in-betweens are allowed. Naturally, not all experts agreed. The following experiment was run with 6 experts and features with 3 possible values. The experts classified 223 Scans. A typical classification vector for one scan and feature is for example {2; 4; 0}. This means that two experts chose the feature value 1 and four experts chose the value 2. The numerical weighted average of this vector is $1/6 * (2 * 1 + 4 * 2) = 1.67$. It has the variance $var = 0.22$. The statistical analysis of the experiment showed an average variance per feature of 0.2, hence the experts go into the same directions.

The variance is not equally distributed among all features. This means not necessarily that some are a more critical (controversial) than others. It could also be the case that the truth lays in between two classification values. No feature has such high values that the classification seems not feasible. Finally, a correlation analysis was needed to validate the concept of ten independent classification features. The correlation factors range from 0 (no correlation, independency) to 0.4 (low—medium correlation). Therefore all the levels are necessary; none are fully dependent on others.

3 Case Based Reasoning for Morphological Classification

The philosophy of CBR follows the human reasoning: instead of formulating complex rules, humans tend to remember similar situations and to derive decisions from that situations (Aamodt and Plaza 1994). This happens when a doctor makes a diagnosis based on similar previous diagnoses with similar symptoms. The same holds for bankers and their credit decisions. The challenge of the morphological classification lies in the plurality of the experts' opinion. If the classifications are substituted by the numerical weighted average, then the classification vector {1; 2; 1} would have the same value as the vector {0; 4; 0} and as the vector {2; 0; 2}. This loss of information is prevented by maintaining the vector throughout the whole CBR-process. The approach follows here the well-known, established CBR-Cycle (Aamodt and Plaza 1994) (Fig. 2).

A case represents a complete episode and comprises problem and solution. The problem is the scan which consists of number of measurement data from the body scanner. The solution to the problem is the morphological classification according to the scheme described above. The CBR system follows the layered approach: each morphotype feature is treated separately. This is reasonable because morphotypes are to some extend independent of each other. The correlation analysis confirmed this assumption. This means for example that for the classification of the arms scans with similar measures related to the arms are retrieved, independent of the waist, legs or shoulder because scans differing in weight or height can still have very similar arms. The similarity between two scans is therefore never calculated as a whole but always with respect to a given feature, answering questions like: *How similar are two scans to each other with respect to the feature Arm Shape?*

Not all measurements are relevant for all features. Thus, each layer has its own similarity model. These models take into account not only direct measurements but also calculated values. Therefore the similarity is based on distances, girths, angles, differences and quotients. Here is an example: the similarity function of the feature "Relationship Shoulder—Hip" comprises the following data:

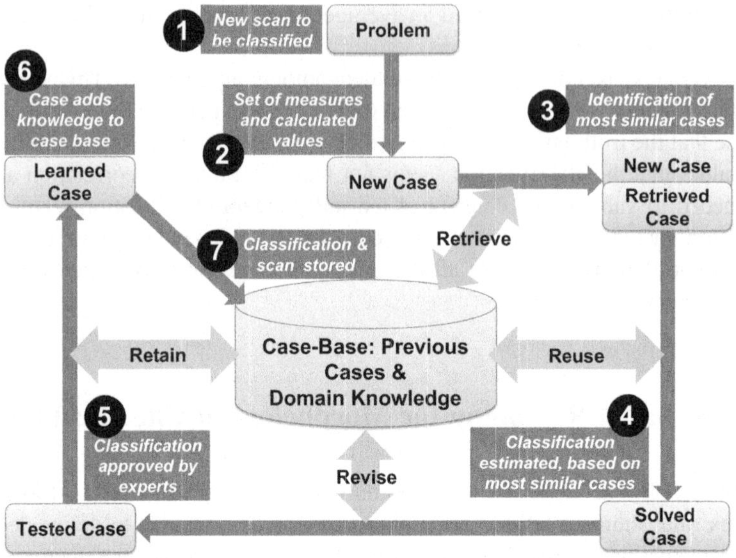

Fig. 2 The CBR-cycle for morphological classification, adapted from Aamodt and Plaza (1994)

Measures: Shoulder Length, Neck Base Diameter, Trunk Length, Hip Girth, Weight, Height, Shoulder Breadth, Hip Breadth
Calculated Values: 2 * Shoulder Length + Neck Base Diameter)/Hip Girth, Shoulder Breadth/Hip Breadth

Not all measures are equally important for a feature. Therefore, the feature model assigns weights to the features. All other classification features have analogue models. The number of data varies between 5 and 25, where some measures like Hip Girth are part of more than one feature model.

The definition of the models is one of the most important knowledge containers of the approach (Lenz et al. 1998). The choice of measures, the definition of calculated values such as quotients and the assignment of weights is crucial for the success of the approach. Therefore, a lot of emphasis was put on the development and fine-tuning of the models.

The similarity between two scans according to the feature-specific models is calculated as the weighted sum of the local similarities between two single measures or calculated data (Lenz et al. 1998). Only numerical, no categorical attributes are used. The similarity as weighted sum thus looks like the following.

Similarity Shoulder Hip(**Case 1**, **Case 2**)
= weight(Shoulder Length) * sim(Shoulder Length (Case 1), Shoulder Length(Case 2))
+ weight(Neck Base Diameter) * sim(Neck Base Diameter(Case 1), Neck Base Diameter(Case 2))
+ weight(Trunk Length)* …

The local similarity can be calculated with a number of different predefined functions available in the CBR-Toolbox MyCBR (2016). For numerical values symmetric polynomial functions seem to work best and are deployed in the system.

Cases with high experimental variance are marked as less relevant due to high uncertainty. The case collection is then the case base for automatic classification of a new data set according to the CBR-approach: it is compared to the data sets of the already classified case base; the most similar data sets are identified and used to propose a classification based on the classification of the human experts.

The RETRIVAL phase of the CBR-Cycle makes use of the described similarity functions and identifies the most similar cases, always with respect to one feature. Based on a number of experiments, the number of retrieved cases was limited to 15 —as long as they have a similarity above 0.75.

The REUSE phase estimates the classification of a new query by using these most similar cases. As the result needs to be a plurality vector again, the system calculates the mean value of the classification vectors of the similar cases. Currently all vectors are weighted equally, modifications are tested at the moment.

The REVISE phase picks up the estimation of the CBR-System and applies it to the problem (query) at hand. This phase is currently in progress. In order to approve and optimize the CBR-Model, auto-classification tests are run. This means that the 223 cases classified by the experts are fed into the system as new problems. The real classification of the experts is then compared to the estimated classification. The test runs show a rather good reproduction of the expert classification. Depending on the given feature, the reproduction rate is around 80%. The values are subject to the quality function applied (mean value versus plurality vector). The auto-classification test runs support the optimization of the similarity functions and the adaptation algorithms.

The RETAIN phase is the last step of the CBR-process. New cases (approved and tested) will be added to the case base if they bring an added value and make the case base learn.

4 Results and Benefits

The described approach has been proved to be valid and comprehensive. It is flexible and extendable because each classification feature has separate similarity and retrieval functions with vast expert knowledge embodied and can be linked to various case-bases. Time consuming individual morphotype classification can be replaced by CBR technology and supports fashion product development as well as size recommendation in online retail.

The most relevant question for the industry is the frequency of morphotype combinations in the different levels. In the following sheet the market shares for women in Plus sizes all features of basic shape in combination with all feature of waist shape are summarized. The different colors mark areas with higher (green) and lower market shares (red). Rectangle is the most frequent basic shape. A consistent distribution of different waist shapes within the basic shapes can be stated. By additional differentiation of age group the changes in waist shape distribution form younger to elderly women can easily be seen. Comparisons of frequency distribution between different sizes show the shift in morphotypes from smaller to larger sizes (Fig. 3).

In the meantime the scientific approach has been evaluated within an industry project for Plus Sizes in der US market with the following results: Different populations with additional shape characteristics need additional classification parameters that can be effectively and easily added to the general classification scheme. All different classifications have to be considered independently due to very low correlations. Significant changes in morphotypes as well as in single classification levels or classifications can be established between different age groups or between different sizes by calculating the frequency distribution in case of market shares. Finally the combined frequency distribution shows the most important morphotypes as a combination of different classifications.

The results of the morphotype classification lead to the following findings: For obtaining higher market shares with a constant number of sizes morphotypes have

		Waist Shape			
		Hourglas	Circle Front	Circle Allover	
Basic Shape		37,16%	32,69%	30,14%	
Triangle	17,13%	6,37%	5,60%	5,16%	17,13%
Rectangle	68,94%	25,62%	22,54%	20,78%	68,94%
Triangle Inverted	13,93%	5,18%	4,56%	4,20%	13,93%
		37,16%	32,69%	30,14%	100,00%

Fig. 3 Distribution of classification, example from industry project

to be regarded (size system development based on morphotypes, differentiated into upper and lower body). The size range and the composition of sizes between the different height types have to consider the different frequency distributions per morphotype to maximize the market shares. If companies split their collection into different height types (petite, regular, tall) the grading needs to be differentiated between the height rows (e.g. different thigh girth grading between petite, regular and tall size rows). Some morphotype classifications change within one size range or height type (petite, regular, tall) so that a non-linear grading is indispensable. Due to different frequency distributions of morphotypes in the different age groups basic block concepts have to be adjusted with respect to the individual target group of the company (Fig. 4).

In the future size system development can be based on morphotype definition and selection so that special basic blocks and fits can be developed with the respect to the most market relevant body shapes. Body measurement portal will allow statistical analysis and size table optimization with respect to single morphotype features or for a target group characterized by a combination of different classification features (Fig. 5).

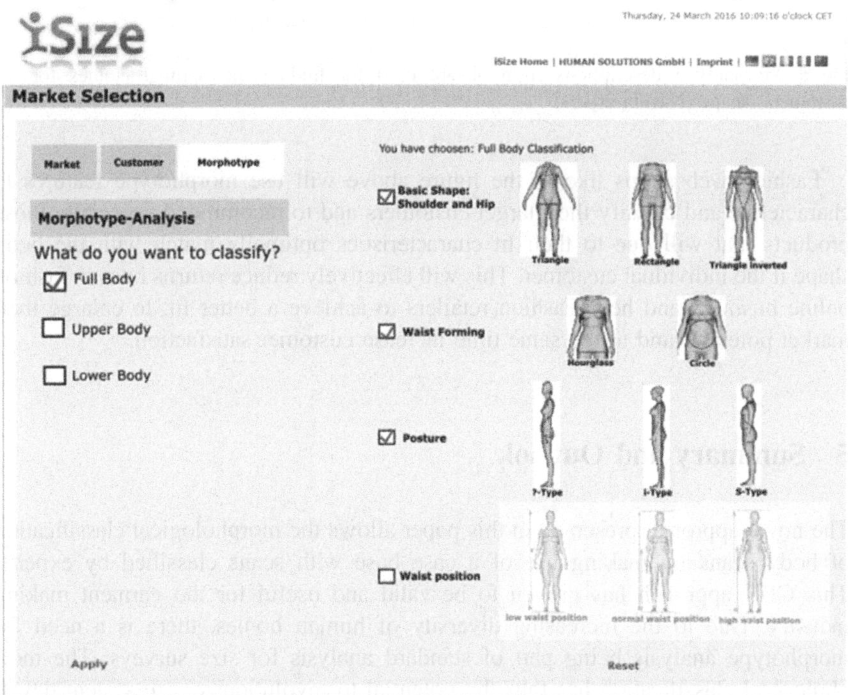

Fig. 4 Example: definition of target customer profile with respect to morphotypes in the iSize body measurement portal from human solutions

Fig. 5 Morphotype descriptions from navabi used for finding the optimal clothes for the customers' shape (Navabi 2016)

Fashion web shops like in the figure above will use morphotype features to characterize and classify their target customers and to recommend especially those products that will due to their fit characteristics optimally match with the body shape if the individual customer. This will effectively reduce returns rates in fashion online business and help Fashion retailers to achieve a better fit, to enlarge their market potential and at the same time increase customer satisfaction.

5 Summary and Outlook

The novel approach presented in this paper allows the morphological classification of body scans by making use of a case base with scans classified by experts. This CBR approach has proven to be valid and useful for the garment making industry. Due to the increasing diversity of human bodies, there is a need for morphotype analysis being part of standard analysis for size surveys. The morphological classification has thus the potential to revolutionize size system development for the fashion industry. Individual customer advertisement by morphotype will be possible.

Acknowledgements The ZIM-Project iMORPH with the grant identification number KF3441501CJ4 is funded by the Federal Ministry of Economics and Technology and runs from 01.02.2015–31.7.2016.

References

Aamodt A, Plaza E (1994) Case-based reasoning: foundational issues, methodological variations, and system approaches. AI Commun 7(1):39–59

Bastos SF, Sabrá Senai Cetiqt FG (2014) The body shape of Brazilian women. In: 5th international conference on 3D body scanning technologies, Lugano, Switzerland, 21–22 October 2014, pp 155–166

Duffy M (1987) HOAX fashion formula, dress the body type you have to look like the body you want. The Body Press, Tucson, Arizona, pp 20–63

Lenz M, Burkhard H-D, Bartsch-Spörl B, Wess S (eds) (1998) Case-based reasoning technology. From foundation to application. Springer, Berlin

Morlock S, Kirchdörfer E, Rupp M, Wendt E, Krzywinski S, Rödel H (2009) Grundsatzuntersuchung zur Konstruktion passformgerechter Bekleidung für Frauen mit starken Figuren. Dresden, Abschlussbericht zu AIF-Forschungsvorhaben, pp 48–53

MyCBR [Internet] (2016). http://mycbr-project.net/. Accessed 24 March 2016

Navabi (2016) Body shapes [Internet]. https://www.navabi.co.uk/fashion-advice/#figurberatungCont. Accessed 24 March 2016

Rissiek A, Trieb R (2012) iSize-implementation of international anthropometric survey results for worldwide sizing and fit optimization in the apparel industry. In: 3rd international conference on 3D body scanning technologies, Lugano, Switzerland, 16–17 October 2012, pp 280–287

Surville J-M, Herichi S, Segalen V (2010) HOAXY body shapes and fashion formula. In: International conference on 3D body scanning technologies, Lugano, Switzerland, 19–20 October 2010, pp 245–258

Webster JM, Cornolo J, Kelkel Y (2012) Comparison of female shape analysis methods for the development of a new sizing system. In: 3rd international conference on 3D body scanning technologies, Lugano, Switzerland, 16–17 October 2012, pp 280–287

GENDE: GENetic DEsign

Best Products Evolve According to Users Feedback

Andrea Vitaletti

Abstract GENDE (http://www.gende.it) is a tool to allow designers, but also common people, to automatically design new products that evolve according to the principles of Genetic Algorithms (GAs). The selection of the products that will actually take part to the evolutionary process, relies on crowdsourcing mechanisms: only the most appreciated products survive. In the era of 3D-printing, GENDE can pave the way to a completely new class of mass products in which personalization become intrinsic to the design process and is driven by common users rather than being confined in the later stages of production and in the hands of professional designers. While GENDE has been originally thought as an automatic design tool, its unique process that involves users from the beginning of the design, can also be used as a powerful marketing tool.

Keywords Interactive design · Genetic algorithms · Crowd sourcing · Participatory design

1 Introduction

GENDE (http://www.gende.it) is a tool to allow designers, but also common people, to automatically design new products that evolve according to the principles of Genetic Algorithms (GAs) (Mitchell 1998).

The process starts with the encoding of the distinctive features of the desired product into a digital chromosome. This allows the generation of an initial random population of products. Such population is evaluated by end-users employing classical online social feedback mechanisms (e.g. Facebook like). The selection of best individuals, relies on crowdsourcing mechanisms, and it is implemented by a fitness function computed over the users feedback on the current population of products. Such best individuals survive and generate a new generation of individuals that

A. Vitaletti (✉)
University of Rome "La Sapienza", DIAG, via Ariosto 25, 00185 Rome, Italy
e-mail: vitaletti@diag.uniroma1.it

R. Rinaldi and R. Bandinelli (eds.), *Business Models and ICT Technologies for the Fashion Supply Chain*, Lecture Notes in Electrical Engineering 413,
DOI 10.1007/978-3-319-48511-9_9

share many of the characteristics of its parents, but can also introduce some mutation. This process is iterated generation after generation until best individuals are positively evaluated by the vast majority of users; the outcomes of the process are 3D models of the best products ready for small-scale series (e.g. 3D print).

Similar ideas have been investigated in generative design (Hartmut et al. 2012; Tedeschi 2014), but mostly in the domain of digital products and without taking fully advantage of GAs. Genetic algorithms have been investigated to generate pictures (Grow your own picture: 2016) or music (Genetic music project. 2016), but only marginally for the collaborative design of physical products (Kram and Weisshaar 2003; Hamda and Schoenauer 2004) and fashion design (Kim and Cho 2005). In the era of 3D-printing, GENDE can pave the way to a completely new class of mass products in which personalization happens from the design process and is driven by common users rather than being confined in the later stages of production and in the hands of professional designers. However, more realistically, in the shorter term GENDE will be employed by product companies to design new products taking advantage of the unique genetic process that naturally embodies marketing aspects (i.e. users feedback) into the design process. In the current stage, GENDE is a proof-of-concept, and still needs some development to be market ready.

Structure of the paper. In Sect. 2 we introduce GENDE concept and we give a short overview on Genetic Algorithms (GAs) principles. In Sect. 3 we discuss the related work and we review some papers that support the use of GENDE not only as a design tool, but also as a marketing tool. In Sect. 4 we present a proof-of-concept of a minimalistic lamp designed by GENDE and finally in Sect. 5, we discuss the main challenges we have to face to make GENDE a real product.

2 GENDE Concept

GENDE is a tool that employs Genetic Algorithms (GAs) for the design of new products. A genetic algorithm (GA), is a search heuristic that mimics the mechanisms of natural selection. In GAs, a population of candidate solutions (named individuals) to an optimization problem evolve toward "better" solutions. Each candidate solution has a set of features that are encoded in a *digital chromosome*. In the example in Fig. 1, the width and length of an eyewear (i.e. an individual) are encoded in the chromosome. A random population of individuals is initially generated. The population evolves in an iterative process generation after generation, as shown in the example in Fig. 2. In each generation, the fitness of every individual in the population is evaluated by a *fitness function*, that in case of GENDE evaluates users' appreciation for the individual, and only the "best" individuals survive and contribute to populate the next generation. In particular, the genome of the best individuals of the current generation are recombined and possibly randomly mutated to form the new generation. Commonly, the algorithm terminates when either a maximum number of generations has been produced, or a satisfactory fitness level has been reached for the population.

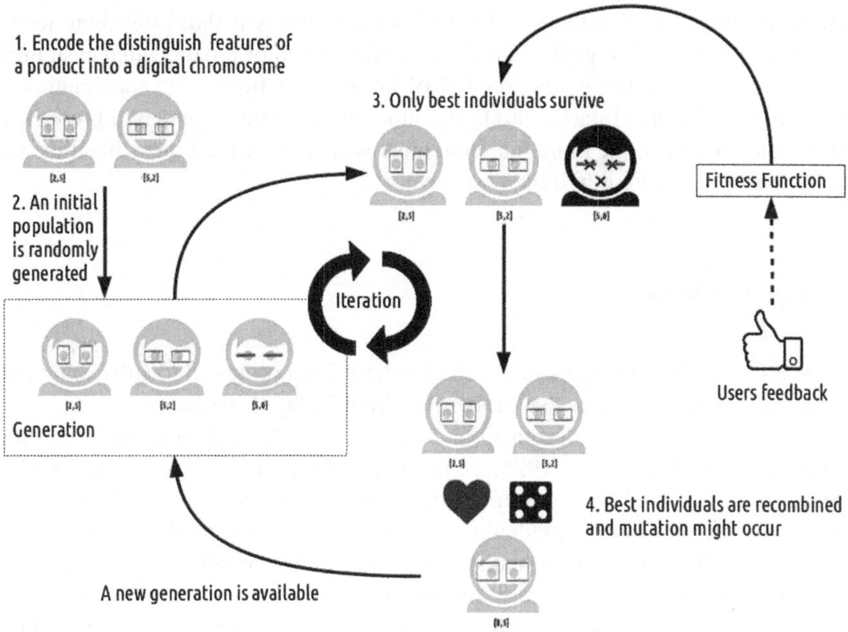

Fig. 1 An schema of the evolution of genetic algorithms

Fig. 2 Three generations of eyewears. In this simple example the fitness function maximizes the size of the eyewear

The application of GAs to the design of new products, requires two fundamental steps that will be further discussed in Sect. 5. First, we have to effectively encode the distinctive features of the product into a digital chromosome. The encoding can be simple in some cases, but in the general case this process is extremely complex. The availability of an effective encoding is the pre-requisite for the generation of the individuals. Secondly, we have to design a suitable fitness function capable to effectively select the individuals that will survive and will thus contribute to the evolution. Since a major goal of the design is the design of products that customers will appreciate, the fitness function of GENDE is a function of the appreciation of users for individuals (Takagi 2001). The more an individual, namely a product, is appreciated by users, the more that individual will score in the fitness function and consequently the more it will likely survive.

3 Related Work

There are a number of papers in which GAs have been proposed to solve a number of design problems, such as the ones presented in Padhye (2012).

A more general product design problem (Rajeev and Ramesh 1987) has been studied in Sundar and Varghese (1996). In this paper, the problem is formulated within the conjoint analysis framework, in which users preferences on a limited number of product profiles are used to predict the valuation for any new product profile which was not initially evaluated. In realistic applications, the number of possible product profiles increases dramatically, and thus the authors propose the use of GAs to explore the product profiles search space. More recently, Vajna et al. proposed the Autogenetic Design Theory (Sndor et al. 2005). In this paper, a 3D parametric CAD is used to create the actual instance of a geometric model to be optimized by a set of parameters and a Finite Element Method (FEM) or a computational fluid dynamics system evaluates the actual instance. The optimization module performs changes on the geometric or topological parameters in order to fulfil the optimization aims. In Ho and Ming (2009) the authors use shape grammars, to express in a general form complex shapes, and GAs to develop stylistically consistent forms applied to the design of a camera. Kram/Weisshaars Breeding Tables project (Kram and Weisshaar 2003), is a renowned example of the employment of GAs to the product design. The algorithm takes in input constraints such as the size, weight, and tabletop height, then generates hundreds of configurations that are selected by the designers. Similarly, the Computational Chair project developed by EZCT Architecture and Design research (Hamda and Schoenauer 2004) uses a GA to generate design variations of a chair. While both these projects rely on GAs, they do not envision as in our case, the involvement of users in the implementation of the fitness function.

The involvement of users is explicitly considered in the interactive evolutionary computation (IEC) where the evolutionary computation optimizes systems based on subjective human evaluation (Takagi 2001). In Zhun et al. (2006) the authors explore the integration of human interaction with evolutionary design. In particular they claim that "*the fitness function should be able to take feedback from designers and customers constantly, enabling it to reflect changing market environments or user preferences*". However, they did not provide examples of such virtuous feedback loop. The work of Kelly and others (Kelly et al. 2011) presents an Interactive Evolutionary Systems (IES) that is used to identify the users' most preferred cola bottle shapes among a set of parametrized shapes. Most similar to our work, in Kim

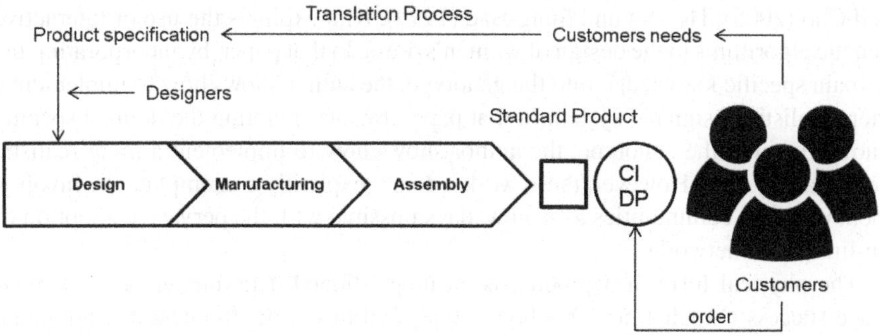

(a) Customers have no additional chance to participate in value creation before the final product.

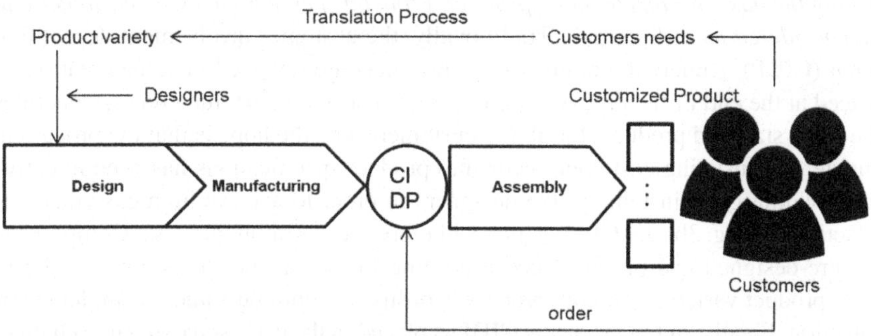

(b) Components are pre-designed and pre-produced according to forecast demands but customization is possible before assembly.

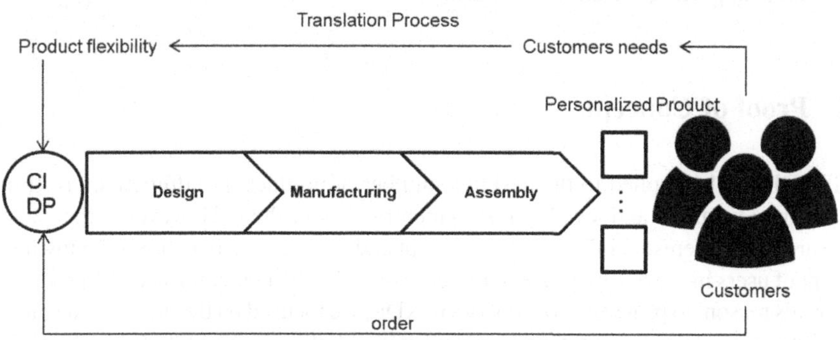

(c) CIDP is placed in the early stage of the value chain to provide customersa real individual customization and their participation since the design phase.

Fig. 3 Customer involvement decoupling point (CIDP) (Risdiyono and Pisut 2011)

and Cho (2005); Hee-Su and Sung-Bae (1999), Kim explores the use of interactive genetic algorithms to the design of women's dress. In that paper, by incorporating the domain specific knowledge into the genotype, the author shows how to implement a more realistic design aid system.In that paper, by incorporating the domain specific knowledge into the genotype, the author shows how to implement a more realistic design aid system. However, these works, do not explicitly contemplate the involvement of users' communities as is nowadays possible with the pervasive adoption of on-line social networks.

The physical form of a product is an unquestioned determinant of its marketplace success (Mitchell and Xuehong 1998) and this is the fundamental argument to support the interconnection of marketing and design, however their connections are not fully explored (Margaret and Lucy 2007). In Risdiyono and Pisut (2011) the authors explore the concept of design by customer: "*Design is a process of establishing the basic parameters of a product. From this perception, every customer can be considered as a designer*". Traditionally, the customer involvement decoupling point (CIDP), namely the point where customers are involved in value creation, is placed at the end of the process (see Fig. 3a). In such context, the process generates the same standard products for all the customers, and the hope is that the process in charge of translating customer needs into product specifications has been effective in providing good insights to the designers in order to achieve high customer satisfaction. In Fig. 3b, the CIDP is placed before the assembly process. Components are pre-designed and pre-produced according to forecast demands but in order to offer product variety, assembly process is postponed until customers complete their selection. Finally, in Fig. 3c a new CIDP is placed in the early stage of value chain to provide customers a real individual customization and support the design by customer (DBC) approach. In this perspective, GENDE can be considered among the first tool supporting the DBC approach.

4 Proof of Concept

GENDE can be applied to the design a number of products in different areas, from fashion to furniture and jewelry, just to mention few of them. However in this paper we are more interested in testing the concept and the process that allows the involvement of users in the design process, rather then in the final outcome (i.e. the product). For this reason, to prove the concept of GENDE, we focused on the design of a simple desk lamp made of four rectangles panels with a hole in the middle (Figs. 4 and 5). The sizes of the holes is the result of the evolution implemented in GENDE. The GAs are implemented in the Python evolutionary computation framework DEAP (https://github.com/DEAP/deap). The population is made of 10 individuals (i.e. lamps), with $cxpb = 0.5$ (the probability of mating two individuals) and $mutpb = 0.2$ (the probability of mutating an individual). Each individual in the population is evaluated by 10 users and the fitness function for an individual is the sum of positive feedback received by that individual (i.e. the appreciation of the users). We stress that the pur-

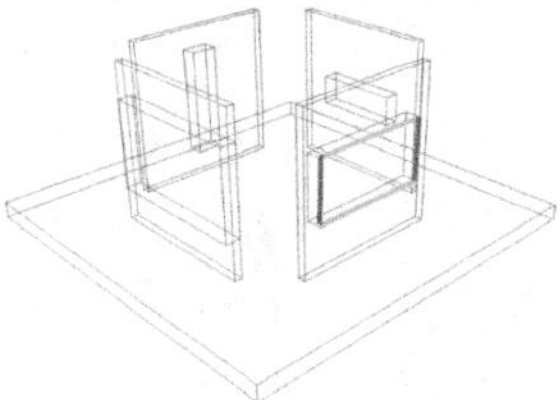

(a) The holes in the lamp are encoded in the chromosome and evolve.

(b) A render of an individual (i.e. a lamp) evaluated by users in the 3rd generation.

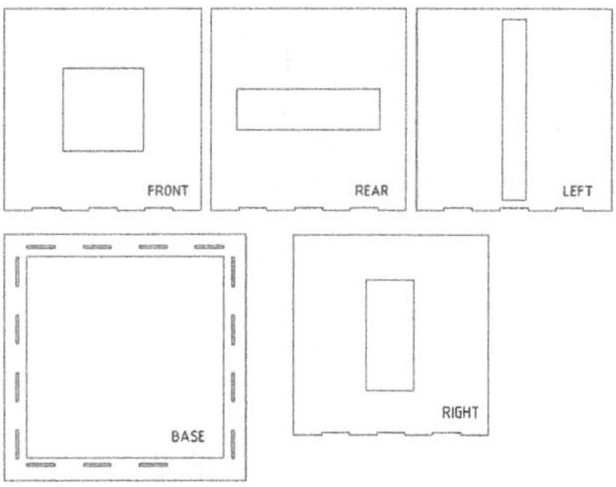

(c) The drawing for the lasercut.

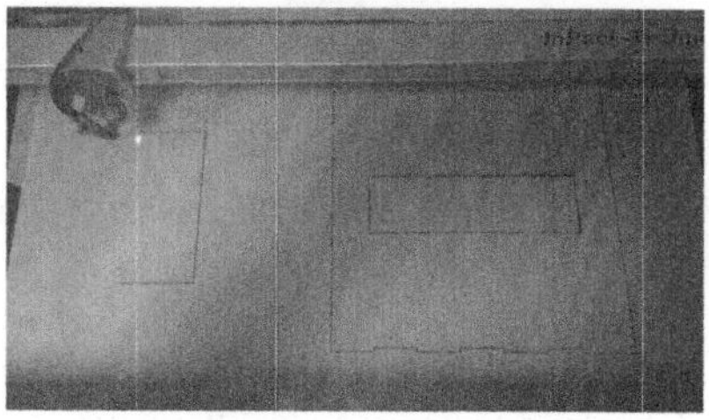

(d) The lasercut of the parts.

Fig. 4 A lamp designed by GENDE

Fig. 5 The final product

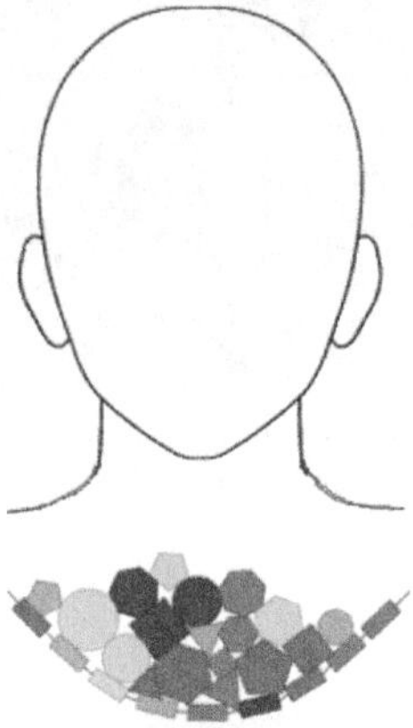

(a) A GENDE necklace.

(b) A t-shirt of an individual in genertaion 11.

Fig. 6 An example of GENDE applied to fashion accessories

pose of this initial experiment is to test the process, rather than the quality of the final outcome that necessarily requires the involvement of a larger number of users and a also a larger population of individuals.

Approaches similar to GENDE have been applied to the design of fashion clothes (Hee-Su and Sung-Bae 1999; Kim and Cho 2005). However, we do believe that fashion accessories, such as handbags, eyewear (see Fig. 2), jewelry and watches, might be more effectively designed by GENDE. In Fig. 6 we show an example of GENDE applied to the design of a necklace. The size, shape and color of the elements in the necklace evolve generation after generation according to users' feedback.

5 Discussion

In this paper we presented GENDE, a design tool based on Genetic Algorithms to support designers, but also common users, in the development of new products. Thanks to its unique participative design process, GENDE is not only a design support tool, but also a valuable marketing tool that brings the Customer Involvement Decoupling Point (CIDP) (Risdiyono and Pisut 2011) in the early stage of the value chain. The Proof of concept presented in Sect. 4 shows that GENDE can be actually used to design simple new products. However, to successfully apply the envisioned process to the design of more complex and general products, we still have to face a number of challenges, some of which are briefly discussed in the following.

The digital chromosome of complex products. When simple shapes are considered, the encoding of the distinguish features of a product into the digital chromosome is relatively simple. However, if the considered product is made of complex and interconnected shapes, their encoding can be difficult and it can lead to very long chromosomes that possibly make the application of GAs challenging both in terms of effectiveness (i.e. converge towards "good" solutions) and efficiency (i.e. converge in relatively small time).

Fitness function and user involvement. The fitness function selects the "best" individuals in a population. If the objective function is clear, objective and well defined (e.g. maximize the size of a shape) the application of GAs is straightforward. However, in our case, we aim at evaluating the appreciation of the users for a product. To this purpose, we use a feedback mechanism based on crowdsourcing; the current generation of products is shown to the users that can simply select via a Web and mobile App, the products they like more. To do not annoy users, only a limited number of products should be shown to each user; the system take care of getting the same number of feedbacks for each individual in the population. An exciting perspective to improve the effectiveness and efficiency of users' evaluation is the employment of neuromarketing techniques (Dan and Gregory 2010), in which users' appreciation can be evaluated analyzing the output of a Brain Computer Interface (BCI).

The fitness function for a given product is evaluated as a weighted function of the positive feedback for that product. To get more significant results, a critical mass of users should be involved, in order to guarantee unbiased results and a relatively fast convergence towards a population of "good" candidate products. The methods envisioned to involve such critical mass of users are beyond the scope of this paper, but foreseen the adoption of social and online marketing strategies.

Design versus marketing. The unique process envisioned in GENDE, guarantees that the generated products are appreciated by users, however cannot not guarantee at all that those products are not ordinary in terms of design.

Citing Don Norman "*Homogenization is disturbing. It diminishes the richness of life, the importance of historical roots, ritual and custom. Cultural diversity is a powerful, positive influence and we, as responsible designers, should pay much attention to how people behave in their environments, supporting the richness of cultural diversity. It is not the product that is important: it is how it is used, in context.*"

In this perspective, GENDE is not only a participatory design tool capable to involve in the design process a mass of users, but also a marketing tool in which users can contribute bringing into the process their own unique needs, stories and backgrounds. It is then again in the responsibility of the designers the possible interpretation of such contribution in the most creative and innovative way, with the purpose of realizing unique products.

Acknowledgements The author wish to thank the FABLAB ROMA http://www.laziofablab.it/, Via Casilina 3/T, 00182 Roma, for their kind support in the production of the prototype in Fig. 5.

References

Ariely D, Berns GS (2010) Neuromarketing: the hope and hype of neuroimaging in business. Nat Rev Neurosci 11(4):284–292

Balakrishnan PVS, Jacob VS (1996) Genetic algorithms for product design. Manage Sci 42(8):1105–1117

Bruce M, Daly L (2007) Design and marketing connections: creating added value. J Mark Manage 23(9–10):929–953

Fan Z, Andreasen M, Wang J, Goodman E, Hein L (2006) Towards an evolvable engineering design framework for interactive computer design support of mechatronic systems. In: Proceedings of the 2006 conference on integrated intelligent systems for engineering design, Amsterdam, The Netherlands. IOS Press, pp. 182–198

Genetic music project (2016). http://www.geneticmusicproject.com/. Accessed 13 July 2016

Grow your own picture: genetic algorithms and generative art (2016). http://chriscummins.cc/s/genetics/. Accessed 13 July 2016

Hamda H, Schoenauer M (2004) Studies on optimization: computational chair design using genetic algorithms. EZCT Architect Des Res

Hartmut B, Benedikt G, Julia L, Lazzeroni C (2012) Visualize, program, and create with processing. Generative design. Princeton Architectural Press

Kelly JC, Wakefield GH, Papalambros PY (2011) Evidence for using interactive genetic algorithms in shape preference assessment. Int J Prod Dev 13(2):168–184

Kim H-S, Cho S-B (1999) Development of an iga-based fashion design aid system with domain specific knowledge. In: 1999 IEEE international conference on systems, man, and cybernetics, 1999. IEEE SMC '99 conference proceedings, vol 3, pp 663–668

Kim H-S, Cho S-B (2005) Fashion design using interactive genetic algorithm with knowledge-based encoding. Springer, Berlin, pp 411–434

Kohli R, Krishnamurti R (1987) A heuristic approach to product design. Manage Sci 33(12):1523–1533

Kram R, Weisshaar C (2003) Breeding tables. http://www.kramweisshaar.com/projects/breedingtables.html

Lee HC, Tang MX (2009) Evolving product form designs using parametric shape grammars integrated with genetic programming. Artif Intell Eng Des Anal Manuf 23:131–158

Mitchell M (1998) An Introduction To Genetic Algorithms. MIT Press, Cambridge

Padhye N (2012) Evolutionary approaches for real world applications in 21st century. In: Proceedings of the 14th annual conference companion on genetic and evolutionary computation, GECCO '12, New York, NY, USA, 2012. ACM, pp 43–48

Risdiyono, Koomsap P (2011) Design by customer: concept and applications. J Intell Manuf (2011) Design by customer: concept and applications. 24(2):295–311

Takagi H (2001) Interactive evolutionary computation: fusion of the capabilities of ec optimization and human evaluation. Proc IEEE 89(9):1275–1296

Tedeschi A (2014) AAD algorithms aided design. Parametric strategies using grasshopper. Le Penseur

Tseng MM, Xuehong Du (1998) Design by customers for mass customization products. CIRP Ann Manuf Technol 47(1):103–106

Vajna S, Clement S, Jordan A, Bercsey T (2005) The autogenetic design theory: an evolutionary view of the design process. J Eng Des 16(4):423–440

Turning a Lean Business Model into a Successful Start-up in the Wearable Technology Sector: The Case of Clara Swiss Tech

Marco Dal Lago, Donatella Corti and Paolo Pedrazzoli

Abstract In the last years the world witnessed a new start-up renaissance driven by the agility of customer-oriented lean business models. Lean start-ups have demonstrated the opportunity to overcome the entry barriers of markets consolidated by large companies. The most innovative aspect of the lean start-up approach compared to one adopted by traditional fashion companies is the focus on the building of a minimum viable product instead of relying on time consuming and unachievable business plans. In this paper the main lean start-up methodologies are presented through a case study of a Swiss start-up operating in the wearable technology field, CLARA Swiss Tech.

Keywords Wearable technologies · Lean canvas · Lean business model · Lean start-up · Smart clothing · E-textile

1 Introduction

The world has changed. In the last few years the market witnessed a real renaissance of the start-up and entrepreneurial eco-system where "three-person" start-ups have grown at such a speed to be nowadays comparable—in terms of revenue—to centenary established enterprises. The table below briefly depicts some key figures of today's disrupting companies both in the product and service sectors.

Many studies describe this revolution as a shift from traditional business strategies to innovative "lean start-up" methodologies (Blank 2013). According to the old and academic formula, entrepreneurs start by writing a business plan from an idea, then they raise investments, set up a team, and develop a final product for the customers. On the contrary, the new lean start-up approach starts from the development of the so-called minimum viable product (MVP) by strongly engaging

M.D. Lago (✉) · D. Corti · P. Pedrazzoli
University of Applied Sciences and Arts of Southern Switzerland,
6928 Manno, Switzerland
e-mail: marco.dallago@supsi.ch

R. Rinaldi and R. Bandinelli (eds.), *Business Models and ICT Technologies for the Fashion Supply Chain*, Lecture Notes in Electrical Engineering 413,
DOI 10.1007/978-3-319-48511-9_10

Table 1 Some figures related to new disrupting start-ups (adapted from different online sources)

	Year of establish	Revenue (R)	Employees (E)	Ratio R/E	Valuation
Whats App (Service)	2009	$ 1 B (2017)	55	18.18 M	19B (2014)
Uber (Service)	2009	$ 498 M (2015)	3000 (HD)	0.16 M	68B (2015)
Airbnb (Service)	2008	$ 340 M (2015)	2368	0.15 M	24B
Pebble Tech (Product)	2012	$ 96 M (2014)	70	1.38 M	–
Oculus (Product)	2012	–	101–250	–	2B (2014)

the customers from its conception. This paper describes how CLARA Swiss Tech Sagl (http://claraswisstech.com), a start-up born in Switzerland in 2015 operating in the Wearable Technology (WT) sector, has successfully adopted this approach (Table 1).

2 CLARA Swiss Tech Sagl

CLARA Swiss Tech is a successful start-up established in Lugano (Switzerland) in September 2015. Everything started in January of the same year when three master students of the University of Applied Sciences and Arts of Southern Switzerland wished to find a solution to save cyclists and pedestrians' lives in the ever-growing city traffic. Exploiting their complementary expertise, they had the idea of integrating LED systems and smart-sensors into fabrics, thus conceptualizing what would have become the CLARA's smart jacket (Fig. 1).

2.1 The Problem

Whilst cycling has beneficial effects both on the health of people and the environment, the on-going increment of city-traffic poses a major threat to cyclists' safety. Every year around 2000 cyclists and 6000 pedestrians are killed in road traffic accidents in Europe (2012). Research shows that one of the major causes of collision with cars is due to poor light conditions.

In fact, although only 10% of bike rides take place at night, around 30% of car-on-bike accidents and 50% of car-on-pedestrian accidents occur during night-time. Despite these alarming numbers up to 40% of cyclists ride in the dark without lights or with partially defective lights. Moreover, a large number of recent studies show that these road accidents occur more frequently in urban areas and close to intersections.

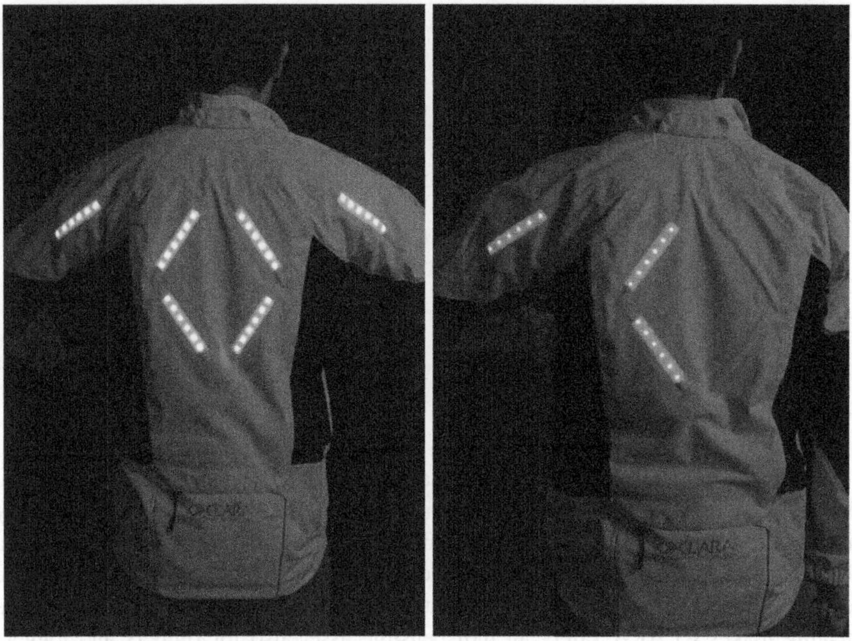

Fig. 1 CLARA's smart jacket: the prototype

2.2 *The Solution*

As a result of an intensive analysis pointing out the main causes that lead to such dangerousness on the city roads for riders and runners, the team invented a smart jacket called CLARA.

> CLARA is a revolutionary smart jacket that incorporates ultra bright turn signals and braking sensors to improve cyclists and pedestrians' safety around the world through enhancing their visibility in poor light conditions (http://claraswisstech.com).

The most innovative component is the brain of CLARA smart jacket, which is a small circuit hidden into a pocket located in the internal part of the jacket. The circuit is covered by polycarbonate material to ensure its quality and waterproofness. The circuit is then connected to a series of flexible LEDs circuits through thin conductive threads heat-sealed in the internal surface of the jacket. The ultra bright LEDs are located onto the back of the jacket forming two arrows for the turn signals, and both on the front and back of the sleeves to guarantee a 360° visibility. The activation mode of the turn signals was made in the best possible way in order to not interfere with the standard user habits. A small lever placed on the handlebar of the bike (see Fig. 2) allows to lightening up the turn signal accordingly to the chosen direction.

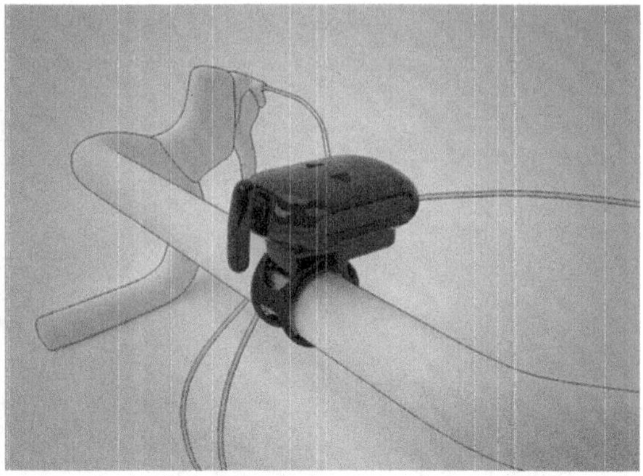

Fig. 2 CLARA handlebar wireless lever

Moreover, CLARA incorporates an accelerometer that perceives whether the rider is braking and automatically activates all the indicators in red to promptly indicate the slowdown to upcoming car-drivers and other road users. Eventually, a button placed directly onto the smart jacket allows the user to lighten up all the embedded LEDs in several different situations, either in a solid or strobe mode. This feature makes CLARA safety an indispensible asset also for runners and joggers who recognize the importance of great visibility in poor light conditions.

The start-up deposited in August 2015 an international patent of the invention and demonstrated its value among several competition:

- MassChallenge UK: top 218 highest-impact start-up globally out of more than 2200 applicants;
- AIT Camp India 2015: top 15 selected Swiss start-up for the program;
- StartCup Ticino: top 5 start-up of Southern Switzerland;
- Wearable Technology Show 2016 (London): Runner Up for the best consumer product award.

The success of CLARA is due not only to the originality of the idea, but also to the lean business model approach used to manage its growth whose analysis is the focus of this paper.

3 Lean Start-Ups and Lean Business Models

Nearly 90% of the start-ups founded every year fail Patel (2015). In fact, there are many potential failure points for start-ups (e.g. hiring wrong people, draining investments, scaling up quickly, building the wrong product, etc.). However, a new approach giving the premises to overcome some of these issues is emerged in the last years: the lean start-up methodology (Furr and Dyer 2014).

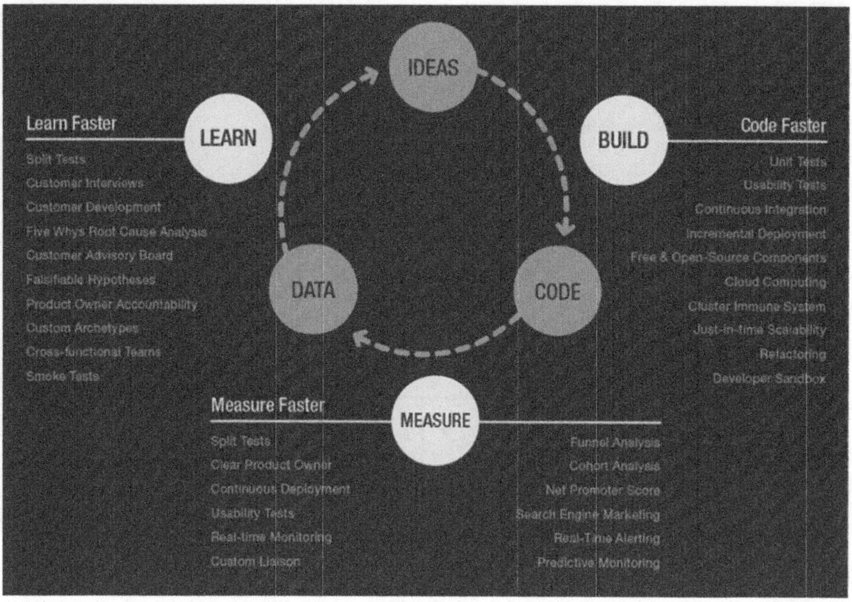

Fig. 3 Lean start-up build-measure-learn process

The US serial entrepreneur Eric Ries, coined the term in his best seller (Reis 2011) relying on the fact that entrepreneurs can strongly reduce the product development time and lower the failure's likelihood through business-hypothesis experimentations and validate learning by engaging early adopters who can foresee the start-up vision since the product ideation stage.

One of the key processes to tackle this challenge within the lean start-up methodology is the build-measure-learn feedback loop (see Fig. 3).

Along this process, the first step is to develop the so-called minimum viable product (MVP). An MVP is characterized by a minimum set of features meant to deploy the product to the customers and gather valuable feedbacks in the earliest stage possible. This method aims at avoiding the manufacturing of products based on assumptions that often customers do not want. As Eric Ries says in its famous lean start-up book (Reis 2011):

> The minimum viable product is that version of a new product which allows a team to collect the maximum amount of validated learning about customers with the least effort.

Once the MVP is established, the start-up can then measure and learn from the customers, who thanks to the deployed MVP, can provide valuable feedbacks for the start-up. This approach leads entrepreneurs to design products that are strictly tailored for the customers and to avoid expensive and time-consuming in-house researches that frequently lead in building the wrong product.

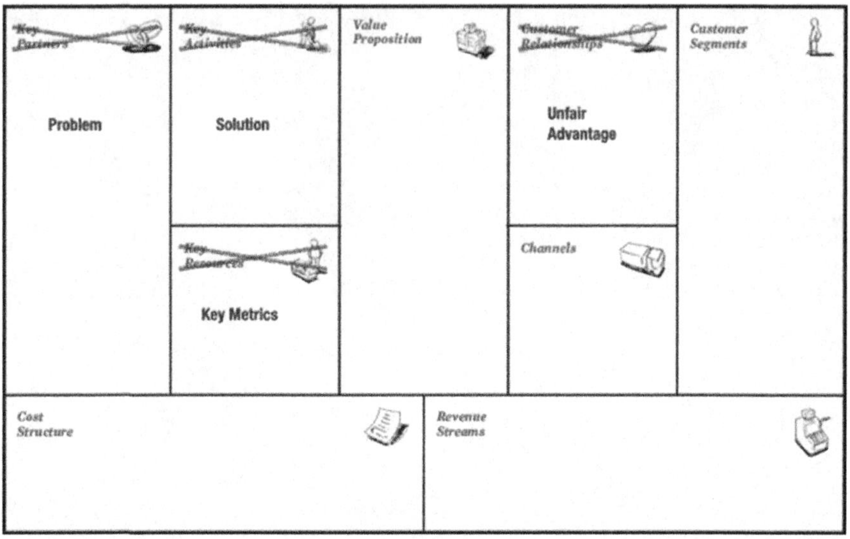

Fig. 4 Template of a lean business model canvas

For this reason, the lean start-up approach brought new pragmatic tools to support entrepreneurs' strategy. On of those is the lean canvas (see Fig. 4), which is an adaptation of the renowned business model canvas created by Osterwalder and Pigneur (2010).

The lean canvas focuses on problems, solutions, key metrics and unfair advantage. Often, lot of resources are wasted in building up the wrong product, so identify the problem is key when designing a product. When the problem is clearly defined, the solution—consisting in the MVP—can be identified. The key metrics box allows the start-up to identify which are its key engines of growth in order to measure the activities that make customers more engaged with the product, those features that the customer is willing to pay for.

Furthermore, often start-ups struggle to find out what their advantage against competitors is. If they reach a good level of success it is inevitable that competitors will enter the market. Therefore, the unfair advantage box helps start-up focusing in their core competences as a defence against competition.

4 From the Idea to a Business Model

CLARA was born from the idea of 3 young students/researchers and keen cyclists of the University of Applied Sciences and Arts of Southern Switzerland.

As the first step, CLARA's team decided to start with prototyping the product from day one rather than writing a business plan. After a brief state of the art analysis looking at the main technologies that could be used for integrating LEDs and circuits into garments, CLARA's team proceeded to design the hardware and software of the smart jacket.

Since the technology was the really innovative part of CLARA, the team decided just to firstly integrate the circuit into a white-label jacket with the help of a local tailor.

This prototype was the real bootstrap of the start-up. Indeed, with the prototype the team was then able to showcase not only the concept but also a tangible realization of the product. Firstly, a short video clip was made in order to present the product on social networks and over the start-up website to stimulate traction and feedbacks from the customers.

Thanks to this engagement with the customer CLARA Swiss Tech Sagl was going further with the prototypes by strictly integrating only the features really valuable for the customers. This was made through several surveys directly to urban riders, focus groups and a dedicated platform (http://www.ideas.claraswisstech.com) where customers could comment or give new ideas about new features to be integrated. Afterwards, other people could vote the comments already placed in order to highlight the most-desired upgrade of the jacket so that the team could evaluate its feasibility and integrate it. As a result, this platform allowed to shape the product in line with the customer needs. For example, an issue that was pointed out was related to the fact that most of cyclists wear a backpack and the lights placed on the back of the jacket might be hidden. In order to overcome this issue, other LEDs were integrated also onto the sleeves of the jacket thus to have a 360° of visibility.

Furthermore, this process resulted also in the development of other products that were requested by the customers: the backpack cover and the gilet.

The product alone was not enough to launch a new business, the team understood after a short period of beta testing of the prototypes by urban riders, that it was the right moment for taking some investments and to start thinking about the industrialization of the product. Thanks to the MVP manufactured the team did not struggle a lot for finding such investments in Switzerland from business angels. This seed money allowed the team to pass between rough prototypes to 0-series products since they could now activate all the suppliers able to make the mass production for CLARA.

Another important step achieved thanks to this first seed investment round was the deposition of an international patent. This led the team to showcase the product to several important events and keep pace for advertisements and consequential opportunities for investments/partnerships.

In this way the team participated successfully in many start-up competitions, events and acceleration program. After being selected as one of the 218 highest-impact start-ups in one of the largest acceleration program worldwide called MassChallenge (London), CLARA Swiss Tech Sagl gained quite huge popularity that led also the start-up to be featured on Forbes magazine (Salter 2015) and to be selected as runner up for the best consumer product award at the wearable technology show 2016.

5 The Lean Business Model Applied to CLARA

The team adopted the lean business model canvas and used it to spread the start-up strategy to many stakeholders (investors, application for competitions, partners, etc.). The lean canvas revealed to be the best method to turn the increasing interest towards the product into a profitable business. In the following, how the lean business model canvas has been applied to CLARA is described by presenting how the blocks have been interpreted.

5.1 Problem

While cycling has beneficial effects both on the health of people and the environment, the on-going increment of city-traffic poses a major threat to cyclists' safety. In fact, research shows that one of the major causes of collision with cars occurs in poor light conditions and close to intersections. This issue can be valuable also for runners and joggers who recognize the importance of great visibility in poor light conditions According to the approach proposed by the lean business model, this block is the one that makes the difference compared to traditional business models. In order to being able to characterize the problem as carefully as possible, CLARA's team explored firstly free data available on the web (cycling forums, blog, etc.) to discover useful information about customers' experience with similar products. What it was found out was that the customers were already trying to develop by themselves this kind of safety clothing but apparently no solution saleable, in terms of certifications, waterproofness and reliability of the technology could be found on the market. Basically, the problem was clearly stated by thousands of potential customers online that were asking for such a product; what was missing was a professional realization and industrialization of it.

5.2 Solution

CLARA is the revolutionary smart jacket that incorporates ultra bright turn signals and braking sensors to improve cyclists and pedestrians' safety around the world through enhancing their visibility in poor light conditions.

5.3 Key Metrics

CLARA Swiss Tech Sagl wants to provide a solution which is, on the one hand, affordable for the customers but, at the same time, featuring good quality to stimulate its adoption. To this end, key measurable metrics are related to customer satisfaction (ratings, new customer per month, etc.). In order to make the product accessible from the customer, one important metric would be to monitor the manufacturing costs since it is the primary element in which CLARA Swiss Tech can act to maintain the price low.

5.4 Unique Value Proposition

CLARA Swiss Tech developed a set of products (smart jacket, backpack cover, gilet) combining top fashion fabrics and best smart and e-textile technologies to improve safety in the urban traffic and avoid accidents.

5.5 Unfair Advantage

Firstly, it's worth to mention that CLARA Swiss Tech deposited a patent on the main features of the technology (e.g. wireless connection between handlebar of the bike and jacket, braking sensor, etc.). When looking at the competitors in the market, one key advantage that emerges from CLARA is the combination of top quality and stylish fabrics thanks to its cooperation with Italian apparel manufacturing companies with more advanced technology due to the Swiss expertise.

5.6 Channels

CLARA's strategy foresees two main distribution channels. Firstly, online channel through e-commerce services is meant to create more traction on the market and generate an initial cashflow. Secondly, more traditional distribution channels will be established (CLARA- > Distributors- > Retailers) given the fact that most of outdoor apparel clothing is today sold through retailers.

5.7 Customer Segments

CLARA smart jacket and its key features (turn signals, braking sensor) were designed specifically for urban riders who ride every day back and forth from work. Besides, CLARA's team decided to add a button on their products to directly lightening up the whole indicators on the jacket (through solid and strobe activation modes) in order to make safer also runners and every pedestrian who perceives the risk related to low visibility in the city traffic. CLARA's target are those people safety/environmentally conscious with particular technological affinity.

5.8 *Cost Structure*

In this phase of development, CLARA main costs are related to the research and development activities with their related tasks (patents, certifications). When the launch will start, in a prospective B2C business model, CLARA will face all the costs related to marketing and distribution (advertisements, logistics, customer acquisition, etc.). Moreover, constantly monitoring manufacturing costs (electronic, plastic, garments) will be of primary importance in order to offer an affordable solution to the end consumer.

5.9 *Revenue Stream*

The revenue stream of CLARA strictly depends upon the distribution channel the products are sold through. In the online channel strategy CLARA will take directly the money from the customers, therefore with less middle-players in the supply chain the margin is expected to be higher. For the e-commerce channel, some added mark-ups due to selling service platforms (amazon, ebay, crowdfunding platforms, etc.) have to be taken into account. On the contrary, for the tradition distributor/retailer channel CLARA will sell the final product directly to the distributor taking into account that more mark-ups will strongly higher the price of the end product.

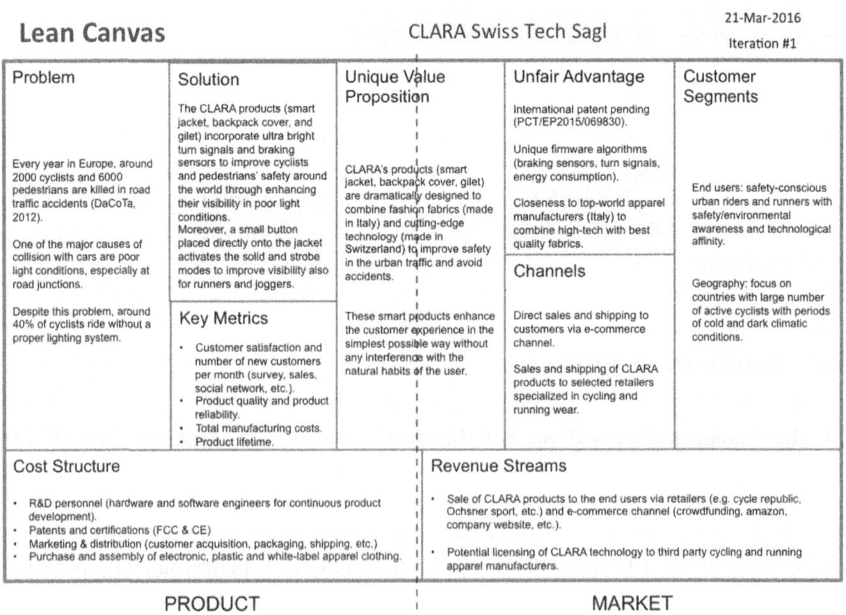

Fig. 5 CLARA lean business model canvas

In Fig. 5 the complete canvas used to foster the CLARA's evolution is presented in detail.

6 Lean Business Model: Application Tips

During the development of the lean canvas the start-up faced also some issues. The main difficulty in building the lean canvas was that the different boxes that compose it have information that usually is available at different time span. For instance, whereas the problem, the solution, and the customer segments were defined quite immediately, some other parts such as the revenue stream and distribution channels were identified only in a later stage, by gaining more insights about the fashion and technology market and after establishing partnerships. Often, there are many different alternatives that can be pursued and they can be more or less difficult to be realized. It will be the choice of the entrepreneurs to select the best coherent option in line with the start-up vision. The selected alternative sometimes could be not the most profitable one, but perhaps, the most likely to be realized both in the short term with the available resources. Another issue encountered from CLARA is that usually the product launch for a start-up takes more time than expected. Several competitors (more if the market sector is on hype) could enter into the market with complementary products and this means that also the unfair advantage and unique value proposition boxes will slightly change from their conception to justify the innovation that the product will have on the market to the stakeholders.

Nevertheless, the lean canvas revealed to be an incomparable tool to communicate the start-up strategy to the main stakeholders. For the CLARA case it has been used routinely as a one-page business-plan that investors, suppliers, competitions' judges could read to understand in few minutes how the start-up wanted to operate. In this way, it strongly facilitated the feedbacks from stakeholders resulting in many reviews of all the aspects of the business model.

In fact, having such a synthesized big picture of the start-up business model allows entrepreneurs to understand how the different parts of it are interrelated with each other and, in particular, allowed ClARA's team to shape a business model made of coherent decisions aligned with the value proposition. A well defined business model can strongly influence the whole start-up strategy.

7 Conclusion

This paper presents the practical application of the lean business model canvas proposed by Reis (Osterwalder and Pigneur 2010) to a recent Swiss start-up proposing a smart jacket that exploits cutting-edge wearable technologies. The start-up and its history are described before presenting the final canvas developed to support the start-up promotion and growth. Some considerations about the main criticalities, on the one hand, and benefits, on the other hand, from the implementation of this recent managerial tool are presented providing some evidence for its use. The main advantage of the tool advocated in literature has been indeed gained also by CLARA Swiss Tech Sagl: entrepreneurs have been supported in developing products specifically tailored to the customer needs from the very beginning. CLARA Swiss Tech is a very young start-up that is still evolving. Yet, validating the learning through beta test sessions where the lead users can give valuable feedbacks to the start-up's team before huge investments and the mass production is set up already proved to be a winning solution. The agility given by this new approach made possible the creation of new successful start-ups worldwide in the last years, and it seems to be a promising approach also for CLARA Swiss Tech.

From the start-up point of view, the journey ahead is still long with the next steps to be undertaken being a set of make or buy decisions for the production of garments and the selection of suppliers. What is already clear is that tools like the business model canvas are highly useful to support the entrepreneurship of the CLARA's founders, in particular to provide them with guidelines for moving the first steps into a dynamic and complex environment where there is no time to be lost in search of the most suitable option or in developing a wrong product that customers do not desire.

References

Blank S (2013) Why the lean start-up changes everything. Harvard business review
http://claraswisstech.com

European Road Safety Observatory (2012) Traffic safety basic facts 2012. Cyclists. http://ec.europa.eu/transport/road_safety/pdf/statistics/dacota/bfs20xx_dacota-swov-cyclists.pdf

Patel N (2015) 90% of startups fail: here's what you need to know about the 10%, Forbes, January 16. http://www.forbes.com/sites/neilpatel/2015/01/16/90-of-startups-will-fail-heres-what-you-need-to-know-about-the-10/#4115622455e1

Furr N Dyer J (2014) The innovator's method—bringing the lean startup into your organization. Harvard Business Review Press

Reis E (2011) The lean startup. The New York Times

Osterwalder A, Pigneur Y (2010) Business model generation. A handbook for visionaries, game changers, and challengers, pp 1–53

Salter P (2015) Five leading startup entrepreneurs graduating from MassChallenge UK, Forbes, October 16. http://www.forbes.com/sites/philipsalter/2015/10/16/five-start-up-entrepreneurs-with-a-shot-at-winning-masschallenges-half-million/#1e751f7f7fd4

Toward a New Fashion Concepts Design Tool: The vMannequin Framework

Paolo Cremonesi, Franca Garzotto, Marco Gribaudo, Pietro Piazzolla and Mauro Iacono

Abstract IT offers significant tools to foster innovation in the fashion industry. Although the products have a low information content, the information content of the overall development and production process may be rich, specially for high end productions, and the same happens for ancillary processes connected to sales and after sales. In this paper we present the vMannequin framework, designed to assist the sales process by means of an interactive computer graphics assistance system that allows a customer to virtually explore, live or remotely, his experience with a chosen outfit, with possibility of customization. vMannequin is based on off the shelf systems, and is then applicable also to the small business segment, providing new ways for business development.

Keywords Fashion concept design · End user development · 3D computer graphics · Clothing animation

P. Cremonesi · F. Garzotto · M. Gribaudo · P. Piazzolla (✉)
Dipartimento di Elettronica, Informazione e Bioingegneria, Politecnico di Milano, via Ponzio 31/32, Milano, Italy
e-mail: pietro.piazzolla@polimi.it

P. Cremonesi
e-mail: paola.cremonesi@polimi.it

F. Garzotto
e-mail: franca.garzotto@polimi.it

M. Gribaudo
e-mail: marco.gribaudo@polimi.it

M. Iacono
Dipartimento di Scienze Politiche, Seconda Università degli Studi di Napoli, viale Ellittico 31, Caserta, Italy
e-mail: mauro.iacono@unina2.it

R. Rinaldi and R. Bandinelli (eds.), *Business Models and ICT Technologies for the Fashion Supply Chain*, Lecture Notes in Electrical Engineering 413,
DOI 10.1007/978-3-319-48511-9_11

1 Introduction

The level and quality of performances offered by modern off the shelf computers is sufficient to allow the implementation of sophisticated, rich interactive 3D graphical applications with limited costs. Moreover, the computer based entertainment market also provides easily usable and cost effective devices for natural sensing of movement and easy gesture based command, thanks to the diffusion of computer games that found into natural interaction (e.g. real time motion capture) their interaction model. We leveraged these factors to design vMannequin, a framework that can support the sales and post sales segments of the life cycle of market products in the fashion field. vMannequin allows a user to build up his outfit and dynamically see the effect on a virtual mannequin tailored on him, also simulating movements, by means of an affordable platform that can be within the budget of a small or medium shop, thus obtaining an economically sustainable virtual fitting room.

The original contribution of this paper is the presentation of a framework that allows a easy implementation of user support data-driven visualization applications for immersive, interactive, real time outfit design, based on off the shelf computing, sensing and actuation hardware, together to its economical evaluation.

The paper is organized as follows: Sect. 2 presents related works; Sect. 3 provides an outlook on the system; Sect. 4 describes a proof of concept; Sect. 5 offers a brief evaluation of the economical impact; conclusions follow in Sect. 6.

2 Related Works

For more than a decade research and enterprise both have shown interest in virtual fitting rooms Protopsaltou et al. (2002). Their combined efforts brought on market many applications that tried to implement this concept, aiming toward different goals and using different technologies. Web applications are usually developed for straight sales-oriented goals. For example, Virtusize Virtusize Fitting Solution (2016) and similar applications are conceived as plug-ins for e-commerce web sites, while others like Fitnect Fitnect (2016) as full web services. When the goal is the engagement of potential customers augmented reality is exploited as in Swivel Swivel Virtual Try-On System (2016), or the enhancement of self-perception using avatars like in triMirror triMirror Virtual Fitting Room (2016). However, since all these systems are commercial products, it is difficult to find any technical details about them and we focus the related work analysis to those papers closely related to ours. Dress dynamics in real time is considered by some authors, like e.g.: Meng et al. (2010), where a physical based approach is used to realize real-time virtual try-on of garments by user interaction. Their approach is different from ours since the intended use of their application is to test dresses as they are designed by professional, moreover they do not specifically address animation of the virtual bodies used. The introduction of the Microsoft Kinect sensor on market enabled novel interaction approaches that are currently under investigation. In Giovanni et al. (2012), the authors use an high definition camera to record the movement of the user, while the Kinect sensor analyze it.

The analysis is then used to compute dynamic dress fitting which is then composed over the camera recoding. Even if the basic installation setting of the application is similar to ours, their approach is not in real-time. Moreover they do not address customization of dresses. The overlapping technique is also exploited in Hauswiesner et al. (2013), differently from our approach that preferred a virtual mannequin for the fitting. More recently, in Gltepe and Gdkbay (2014) authors exploit the use of the virtual avatars for the fitting. This work lacks the visual appeal that is one of the focus of our proposed system, but introduced the use of real-time virtual body animation. Differently from it, however, we preferred an approach based on the recognition of a movement that triggers the closely matching animation present in a database of animation, instead of matching the user on time. In this way we avoided the animation artifacts that influenced their work.

3 The System

The system we present in this paper is intended to support end-users in developing their fashion concept design. In the following we refer to it as vMannequin. It provides a virtual 3D model, male or female as chosen by the user, that can be dressed as a normal mannequin would be. A variety of 3D assets, ranging from dresses to props, from shoes to hair styles that can be easily but thoroughly customized is provided for the fitting operations. The 3D model is displayed on a big screen, animated in real-time and can mimic users movements. To enhance the realism of the simulation, a great care has been devoted in implementing the visual part of the application using state-of-the-art shading techniques. Dress dynamics is also considered for the same reason.

In this Section we present vMannequin high-level architecture, highlighting all the elements that compose it. The system is shown in Fig. 1. Its components, along with their inter-relationships, will be detailed in the next starting from the input devices that allow users interaction.

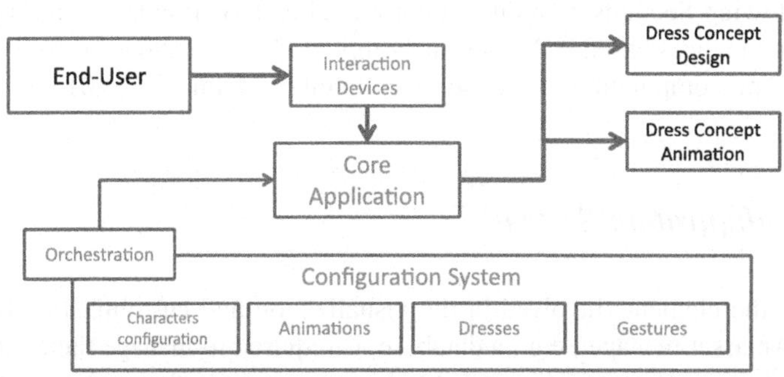

Fig. 1 The simulator structure

3.1 Interaction Devices

Interaction with the system occurs in two forms. The first is *customization interaction,* that is the interaction to select and customize the dresses before fitting them on the 3D character. The accuracy requirement of this interaction form may be high when customization is involved, since it can be a potentially long process with an equally long sequence of gestures to be performed. The second form concerns *animation interaction*, that is the sequence of users gestures that are used to animate the virtual character. Different currently on-market interaction devices can be adopted to these ends, ranging from QR-code reader or sound and motion recognition sensors to smart displays. We group them all under the definition of 'interaction devices' to focus on their ability to capture inputs from users and decode them into parametrized triggers for the application.

3.2 Core Application

The design of dress concepts and the animation in real time of the result are both enabled by the Core Application. It handles the following tasks: render the image on the screen device, load and unload the assets from the database, allow the customization of these assets, react to the inputs received from interaction devices, play the animations. The rendering task requires a good trade-off between performances versus realism. One of the key features of the vMannequin system is its ability to involve potential customers, hence the need for state-of-the-art visual quality coupled with an immediate reaction to user's gestures. Visual quality can be achieved easily nowadays but it usually has an high requirement in terms of per asset storage space, which in turn translates in a possibly higher loading time before the element is fetched from the storage, read and then displayed on screen. To this end the integration with the database is a critical issue, especially when considering precomputed dynamic dresses loading time. We addressed this non-trivial challenge by producing a proprietary binary file format which allowed, along with other techniques described in the following Sections, to reduce it considerably. In particular we used principal component analysis based techniques to compress the animations, as well as a near-exhaustive precomputation of secondary cloth effects Kim et al. (2013).

3.3 Configuration System

Many are the elements involved in the visualization and customization of the 3D models. An asset manager, e.g.: a database, is required to manage them all. Even if this part of the application is intended to be transparent to the final user, it requires careful planning from the developers perspective. The configuration of all the assets that a system like vMannequin proposes to its users requires not only programmers and IT technicians, but also 3D and computer graphics artists, as well as fashion and design experts which rarely have sufficient coding skills. This lead us to develop the Core Application as *data-driven*. The goal is to have a system that can be completely configured by inserting proper information in the database only, in terms of types of assets available, interaction devices used and interaction models. This is to allow the maximum deployment flexibility in different kinds of fashion retail shops. After all, an high-end fashion store has different requirements in terms of customization options, animations, gestures to be recognized than a sport store.

The elements, or assets, required by vMannequin can be divided into four broad categories: Characters configurations, Animations, Gestures and Dresses.

- *Characters configurations*. The virtual models used for dress fitting are essentially one male and one female 3D virtual characters that can be adapted to the user's needs. Features like weight, height, age, eye colors, skin tone, tattoos, nails colors can be customized as required. The database will only have to store the associated parameters, not different character versions. Depending on the degree of customization allowed, this category of assets can require a larger or a smaller storage space.
- *Gestures*. The interaction with the application is governed by gestures. Different type of interaction devices can allow different types of interactions. Full body sensors can recognize the position of the user, and return the orientations and the positions of the various joints that corresponds to the head, the torso, the arms and the legs. They can also recognize a sequence of movements as a specific action performed by the user. Since the application might be configured to be used in different contexts (i.e. a sport goodies store, a bride-dress manufacturer, a department store), different gestures might be required (a sport store customer might want to run or dance, while a bride-to-be might want to throw a bucket). The gesture data-base holds the specific gestures for the considered configuration. These gesture might trigger animation and configuration steps.
- *Animations*. Virtual models are not static objects but can be animated. There are different approaches about their animations. It is possible to map the user's gestures from the interaction device sensor as detected directly to virtual character, but we avoided this approach because it may result in visually awkward movements, that may break the simulation realism. Instead we propose an indirect mapping between the interaction device sensor readings and the animation played. The sensor recognizes an input pattern made by the user and the application select which of the database's available animations is the closest match. In this way, user's actions still affects the virtual mannequin in real-time but the realism of the movement is not compromised.
- *Dresses*. As expected, in a fashion oriented application, this category composes the majority of the database, and its size depends again on the number of dresses, props, hairs the application is required to use. The meshes, i.e. the geometric definition of the 3D objects, can be further divided between *conforming* or *dynamic*. To the first group belong those dresses which follows skin-tightly the 3D virtual

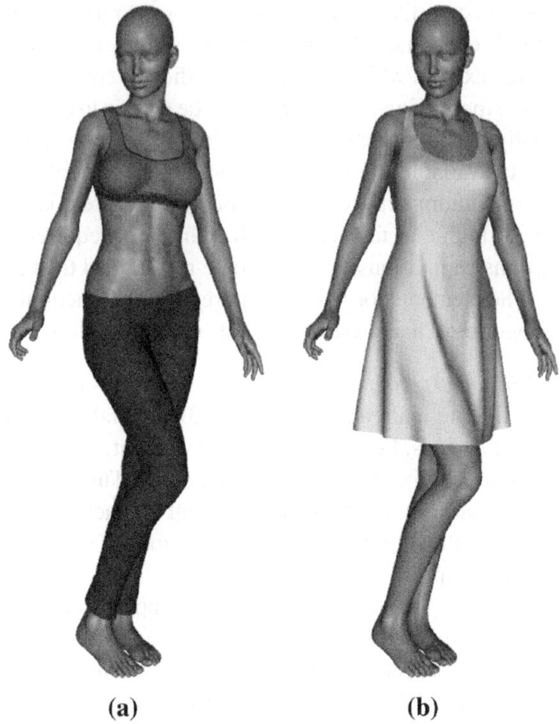

Fig. 2 Character with a conforming (**a**) and a dynamic (**b**) dress

models (such as a pair of leggings). From a computational perspective they are the less expensive to handle because require the same animation techniques used for the virtual characters. Dynamic dresses, on the other hand, try to reproduce the majority of clothes physical characteristics and thus require specific management techniques, often resulting in an higher storage space requirement. In Fig. 2 an example of the two kind of dresses is presented.

3.4 Orchestration

This component handle inter-category communication that is required when assets belonging to different categories need information from each other. Orchestration data holds the information required to connect the gestures, to the dress selection and configuration, to the characters and to the animation. Since the goal is to simplify the customization of the application using a data-driven approach, it exploits a formal specification (described in Cremonesi et al. (2016)) that allows the setup of the interaction model without the requirement of programming skills.

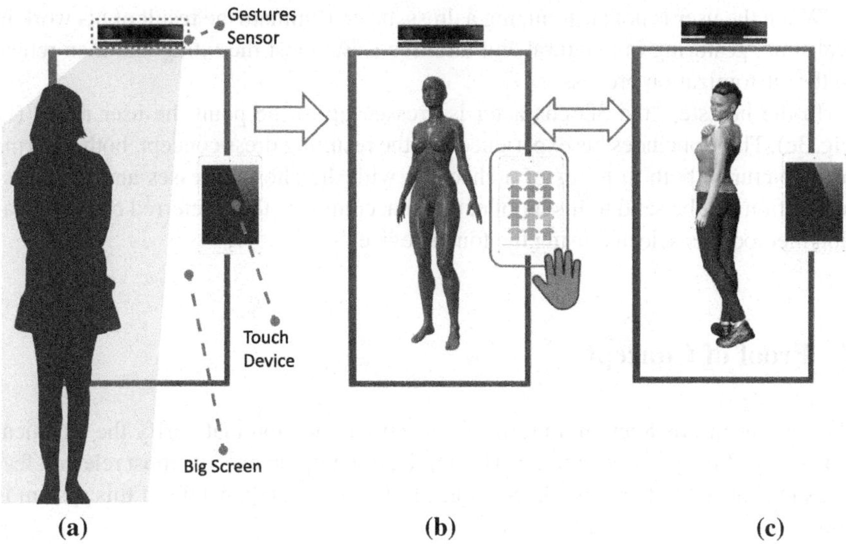

Fig. 3 The user experience. The Kinect sensor detects a user standing in front of the installation (**a**). The user interacts with the application by touch screen (**b**). The user can animate the character fitting of the dress concept created (**c**)

3.5 Users Perspective

In this Section we present a standard configuration of vMannequin, as installed in a fashion store, in order to present it from a user experience perspective. In Fig. 3 the main interaction stages are summarized.

In this standard configuration, we consider the system as composed by two interaction devices. The first one is a sensor capable of detecting the full-body gestures of the user, the second one is a touch device for dress selection and customization. The user can see the dress concept he is composing, in real time, on a big screen.

Once the system's sensor detects the presence of a user, it exits its idle status and becomes active (In Fig. 3a) providing a short set of instructions about the basics of its expected interactions. The user engaged by the system is prompted by the request to specify some characteristics of the virtual character that will help him in designing the dress concept. Features like gender, height, size, age can be specified. This step can be skipped by exploiting automatic user's features recognition methods, but this may introduce more problems than not. What if the dress that the user wants to virtually fit is not meant for him but as a present for someone else?

In the next step a catalog of the available customizable assets (mainly dresses, but shoes, hair styles are available too) is presented to the user (In Fig. 3b). The user is supposed to interact with the application for as long as required to be satisfied with the options selected. This customization process occurs by mean of the touch device. Each time an asset is completed, it can be sent to the 3D character and seen fitted.

When the user is not customizing a dress, he can animate the result of his work in real-time, gesturing in a natural and intuitive way to test the fitting and then return to the customization process.

In the last step, the 3D character is dressed up to the point the user needs (In Fig. 3c). This concludes his experience and the resulting dress concept, both in terms of its pictures, both in terms of a checklist with the chosen dresses and their customization can be send to his mobile, email account or other preferred communication methods, as selected using the touch device.

4 Proof of Concept

We present in this Section a proof-of-concept application that verify the technical feasibility of the proposed model. The application implements the most relevant features introduced in the previous Sections. In Fig. 4 the architecture of this system is shown.

The application has been developed using Microsoft Visual C# (2015), while users can interact with the simulator by means of a Microsoft Kinect II for XBOX ONE sensor Kinect for windows software development kit (2016). The sensor offers a wide range of detection features that can be used to implement both the customization and the animation interaction types, at least for the proof-of-concept installation.

A rendering engine has been specifically developed for the application, tailored on the specific requirements of the simulator. The engine uses a simple but effective three points light model and is capable of a good tradeoff between performances vs realism. To allow the maximum deployment flexibility on different hardware, the OpenGL graphic library has been chosen, accessed through the OpenTK OpenTK Toolkit Library (2016) wrapper. Since it is widely supported by different display adapters, the use of OpenGL could be of benefit especially in case of a future integration with mobile displays.

The critical part of any systems like vMannequin is the quality of models for the asset manager. For the proof-of-concept application we leveraged mainly the free models available for Daz 3D, a software dedicated to morphing, posing, animating virtual characters. Since 3D models can be coded using very different file formats, to

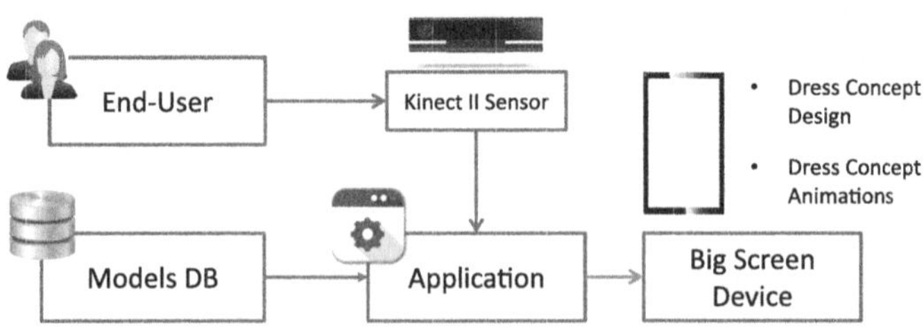

Fig. 4 The system proof-of-concept

avoid forcing the use of a specific one for our application, we used the Assimp Gessler et al. (2016) library, able to import in a uniform manner a vast number of standard formats. Since Assimp is currently unable to read dynamic dress information, stored as mesh *morphs targets* (or *blend shapes*) that only two file format are able to save (i.e.: Collada `*.dae` and autodesk `*.fbx`), we used Autodesk FBX SDK 2015.1 Fbx data exchange technology (2016) to import them. Since performances are an important issue for the concern of assets management (the user should experience immediate visualization of the chosen clothes and the applied customizations), we developed a proprietary binary file format to store them that allowed us to significantly shorten the loading time of dresses, especially of the precomputed dynamic ones. 3D Model files decoded using the above libraries are transformed in the proprietary format used by our application, off-line, before being used by vMannequin system. The animations used in the example application were baked at 30fps, and the simulator can run them while displaying one animating character in full attire, with at least one dynamic dress fit on it. The test was run on a machine equipped with an Intel Core i7 2.4GHz, 8GB of RAM and a NVIDIA GeForce GT670M at a 1920×1080 resolution.

5 Economic Impact Evaluation

The success of an application like vMannequin relies strongly on the quality and number of the asset used during the user experience, as well as the user's perceived interaction quality and the engagement provided. In this section we discuss the main economic issues that may be raised by adopting our solution.

The main aspects that needs to be considered concern the creation of graphical assets. In particular they can be divided into: (i) character creation, (ii) dress creation, (iii) animation. Character creation, to achieve the top quality, requires expert modeler, capable of producing a realistic mesh with vertices and connections carefully placed to obtain good animations. The mesh must support a coherent UV mapping to allow textures to be placed on the skin in such a way to avoid distortion. Good textures should be provided, not only for the main color, but also for transparency, normal and bump mapping, and specular reflection. The character then must be rigged with a skeleton that conform to a prefixed convention, in order to simplify the dynamic animation of the character in the application environment. Such skills are quite rare to find, and real expert might be expensive, making the cost to create a character in the 2–20 K range. The same skills are also required for dress creation. In this case however, two different direction must be followed depending on whether the cloth is conforming or dynamic. In the former case, a rigging process similar to the one used for characters should be followed, otherwise a more physical approach should be followed. In particular, dynamic clothes emulates the dynamic using spring systems whose corresponding coefficients should be carefully parametrized to achieve a result that is coherent with the real dress that is being simulated. Cloth simulation skills are much rarer to find than character creation and rigging, and dynamic cloth parametrization can be really expensive. However, the amount of work required is much smaller: a real expert could set the parameters in very few hours, making the investment in the 300–302 K range. Character animation is instead a well established skill, where several professionals are available, and good results can be obtained even at moderate price range. Moreover the animation cost is proportional to the animation length: this allows the production to choose a reasonable amount of movement to provide good quality results with the available budget. To have a rough idea, the animation cost could range in the 50–300/s. In order to reduce such costs, the best solution is to avoid (or reduce to the essential) the production of completely new assets. The general idea is to use easy-to-find detailed objects that can be near as much as possible the clothes that the retail store wants to show with the application. With a little extra effort by modeling professionals, those objects can be customized to exactly match the clothes in stock. Currently there are different similar programs, e.g. Poser, Daz 3d, MakeHuman, so there is a high number of available dress models, hair styles, props and other elements that can be reused. In particular, careful use of transparent textures can provide slightly different "cuts" to the dresses, allowing a high level of customization using only well known image manipulation software such as Adobe Photoshop or GIMP.

6 Conclusions

In this paper we presented the current state of development of vMannequin, a computer application designed to help fashion store customers to design the dress concept of their next purchase. Future works goes in the direction of improving the performances of the application, also to adapt it to low end architectures. Moreover, we aim to enhance the realism of the simulated environment by means of techniques like spherical harmonics, that allows better environmental lighting of the retail store and better illumination of the 3D models, and achieve an higher degree of realism and immersiveness. Next development stage will be directly tested by users to improve the interaction part, as well as its ability to improve retail store attractiveness.

References

C# Reference for Visual Studio (2015). https://msdn.microsoft.com/en-us/library/618ayhy6.aspx

Cremonesi P, Garzotto F, Gribaudo M, Iacono M, Piazzolla P (2016) vMannequin: a fashion store concept design tool. In: Proceedings by European council for modeling and simulation (ECMS2016)

FBX data exchange technology (2016). http://www.autodesk.com/products/fbx/overview

Fitnect 3D fitting room system (2016). http://www.fitnect.hu/

Gessler A, Schulze T, Kulling K (2016). http://assimp.sourceforge.net/main_doc.html

Giovanni S, Choi YC, Huang J, Khoo ET, Yin K (2012) Virtual Try-on using kinect and HD camera. Motion in games: 5th international conference, MIG 2012, Rennes, France, November 15–17, 2012, Proceedings. Springer, Berlin, pp 55–65

Gltepe U, Gdkbay U (2014) Real-time virtual fitting with body measurement and motion smoothing. Comput Graph 43:31–43

Hauswiesner S, Straka M, Reitmayr G (2013) Virtual try-on through image-based rendering. IEEE Trans Vis Comput Graph 19(9):1552–1565

Kim D, Koh W, Narain R, Fatahalian K, Treuille A, OB́rien JF (2013) Near-exhaustive precomputation of secondary cloth effects. ACM Trans Graph 32(4); 87:1–7

Kinect for windows software development kit (SDK) 2.0 (2016). https://dev.windows.com/en-us/kinect

Meng Y, Mok PY, Jin X (2010) Interactive virtual try-on clothing design systems. Comput Aided Des 42(4):310–321

OpenTK toolkit library (2016). http://www.opentk.com/

Protopsaltou D, Luible C, Arevalo M, Magnenat-Thalmann N (2002) A body and garment creation method for an internet based virtual fitting room. Advances in modelling, animation and rendering. Springer, London, pp 105–122

Swivel virtual try-on system (2016) http://www.facecake.com/swivel/

triMirror virtual fitting room (2016) http://www.trimirror.com/en/about/

Virtusize fitting solution (2016) http://www.virtusize.com/site/

Smart Wearable Multi-sided Fashion Product Platforms

Sergey Yablonsky

Abstract This paper reports on a study of the taxonomy of innovations for wearable products and services. The analysis and discussion lead to a multidimensional framework of innovations, with a particular emphasis on a technology stack, business models, products, services, and platform innovations. We argue that the multi-sided platform might possibly be a successful business model for the adoption of IOT wearable products. We develop IOT-wearable, business-model pattern and the general definition of the IOT- Wearable, Multi-sided Platform. The research contains guidelines to help practitioners and policy makers develop platform-enabled wearable, innovation strategies through the consideration of various levels of wearable business models and MSP strategy. It offers a relevant source of ideas and guidance for anyone interested in research and practice related to rethinking wearable-fashion, product and service innovation.

Keywords IoT · Internet of things · Wearable · Business models · Multi-sided platforms

1 Introduction

Today IT is more and more embedded in products themselves and together with global-network infrastructure forms the Internet of Thigs (henceforth, IoT), transforming value creation through products while triggering a new wave of shifts in the value chain, product design, procurement, marketing, and service while creating the need for new activities.

The IoT device mesh is dynamic and pervasive. Gartner (2014) forecasts that the IoT will reach 26 billion units by 2020, up from 0.9 billion in 2009, and will impact the information available to supply-chain partners and how the supply chain

S. Yablonsky (✉)
Graduate School of Management, St. Petersburg University,
Russia Volkhovskiy Per. 3, St. Petersburg 199004, Russia
e-mail: yablonsy@gsom.pu.ru; yablonsky.serge@gmail.com

R. Rinaldi and R. Bandinelli (eds.), *Business Models and ICT Technologies for the Fashion Supply Chain*, Lecture Notes in Electrical Engineering 413,
DOI 10.1007/978-3-319-48511-9_12

operates. All these billions of things connected with businesses and 3 billion people with smartphones, smart watches, etc. generate vast amounts of rich data, and what companies do with that data–how they turn it into proprietary algorithms and business value–will determine how well they maximize the opportunity presented by digital business.

In order to unlock the potential of IoT, organizations have to master new business models, technology architectures, operating systems, tools, methodologies, databases, networks, middleware, and sourcing partners. Manufacturers can begin now to define their target manufacturing model and then plan the IoT transformation roadmap.

The Internet of Things has been defined (Y.2060 2012) as "a global infrastructure for the information society, enabling advanced services by interconnecting (physical and virtual) things based on existing and evolving interoperable information and communication technologies". These physical and virtual things or smart things could include any object or smart device, such as cloths, a watch, a suitcase, an animal, or even a person. Wearable technology devices, or simply wearables, refer to smart things that are designed to be comfortably worn on the human body. LexInnova (2015) defines: "Wearable technology tends to provide sensory and scanning capability, such as bio-feedback and tracking of physiological function. Wearables also have communication capability which allows humans to access data in real-time using another connected device or medium. Rapidly evolving examples of wearable devices include smart watches, intelligent eyewear, bio-sensing contact lenses, e-clothing, and smart jewelry, such as rings, bracelets, and hearing aid-like devices that are designed to look like ear rings. In some cases, wearable devices may also be implanted into the human body."

Wearable devices and their supportive complementary services have been undergoing anticipation, hype, and speculation mostly during the last decade (Gartner 2015, 2016a, b, c). IoT has the potential to transform the fashion industry and the way we live and work (Gartner 2015). The wearable, smart-devices industry is booming and is being promulgated by the market recently as an alternative to the post-smart-phone industry. It also has unique characteristics of an industrial ecosystem different from the ecosystem of the smart- phone industry.

We analyze the characteristics of wearable applications for IoT scenarios of smart wearables and related products like e-textile, smartwatches (e.g., Samsung's Galaxy Gear, Pebble watch and Qualcomm's TOQ), wristbands (Polar Loop, Nike Fuel band), wearable bio-monitors, etc. Long-term engagement and consumer adoption in the wearable-technology (henceforth, WT) industry are not yet entirely established,; additionally, several companies strive to understand this field of technology in order to create strategic competitive advantage and innovative business models, as well as sustainable and unique value. Much attention is drawn to the efforts by large companies such as IBM, Microsoft, Google, Cisco, GE, Bosch, and government initiatives like Industry 4.0 in Germany (IBM 2015; Microsoft 2015; Google 2015; Cisco 2015; GE 2015; Bosch 2015; Industrie 2013).

Our goal is to conceptualize the IoT, wearable-products technology stack with a particular emphasis on business models, products, services, and platform innovations. We illustrate, analyze, and classify such conceptualization by means of ontologies, in particular lightweight ontologies—taxonomies (Poli et al. 2010; Yablonsky 2014, 2016a, b).

2 The IoT Wearable Innovation Taxonomy

The IoT and related wearables form one of the most data-rich environments on the Web today. A clear and precise description and structuring of the information in the wearables domain are prerequisites for a common understanding of the information exchanged among different partners of IoT/wearables ecosystem. It would also foster semantic interoperability in the context of pan-European data exchange among public administrations, facilitate electronic interoperability among IoT/wearables businesses, and improve statistics on IoT.

In our research, we argue that, rather than searching for a single acceptable definition, a better approach would be to develop a classification system or taxonomy (Poli et al. 2010). Taxonomies and other types of controlled vocabularies are the preferred means to achieve such a common understanding by specifying the terms of the domain, disambiguating them from each other, controlling synonyms, and structuring the domain via term relationships (Yablonsky 2014, 2016a, b). For conceptual grounding of the categories in taxonomy, we use definitions from various information resources. There are many approaches to the classification of the main entities. The current IoT standards activities include several drafts (IEEE-SA 2015; IDC 2015; Taxonomy 2016; 451 Research 2015; MacGillivray et al. 2015). However, access to this data is often inhibited by the lack of adoption of standards by the industry.

The conceptual research presented in this paper aims to collect and analyze quality data regarding the current status and prospective evolution of the IoT wearables domain.

The analysis proposes classification and examination of the current status of wearable technology and services both in industry and academia. The main output is the development of the set of taxonomies in the IoT domain.

A few points should be noted concerning methodology. It includes the method in which knowledge was extracted from a particular domain, the classification and organization of domain concepts, validation issues, and development tools. An appropriate knowledge source consists of academic documents that constitute the entire knowledge domain (e.g., Ugur 2013; Chana et al. 2012; Díaz et al. 2016; Fleisch 2010; Fleisch et al. 2014; Chan 2015; Experton 2016; Green et al. 2014; Hao and Helo 2015; Hui 2014; Hussain et al. 2014; Lee and Lee 2015; Puustjärvi and Puustjärvi 2015; Miorandia et al. 2015; Porter and Heppelmann 2014; Stock and Seliger 2016), a set of industry documents (IBM 2015; Microsoft 2015; Google 2015; Cisco 2015; GE 2015; Bosch 2015; Industrie 2013; IEEE-SA 2015; IDC 2015; Taxonomy 2016, Gartner 2015, 2016a, b, c), reported and observed trends and IoT activity from 2010 to 2016, IoT and wearable technology briefings, press

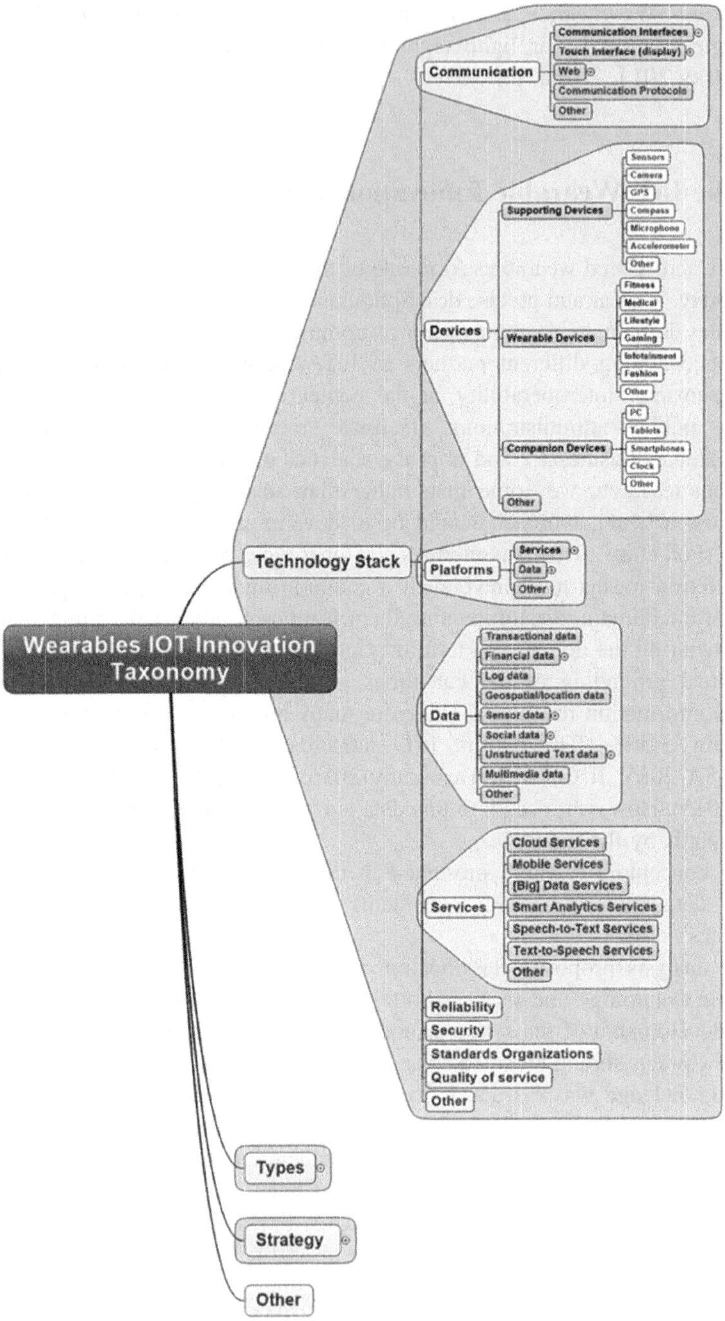

Fig. 1 The IOT wearable innovation taxonomy (technology stack)

releases, market and company's reports, and other publicly available information; web documents (blogs, web conferences, etc.) and market and company's reports.

On the other hand, depending on the nature of the knowledge domain and the taxonomy's purpose, case analysis of source documents and the validation of the first pilot version of the taxonomy by experts in the field can result in superior identification and organization of concepts.

The taxonomy is intended to provide a framework to categorize and relate technology and industry-specific aspects of IoT research area and the market.

The pilot version of the taxonomy (hierarchy of some taxonomy concepts—is shown in Figs. 1, 2 and 3; a set of instances—international and local IoT and fashion companies, services, etc.) contains more than 300 terms; the total number of all taxonomy features is 400. So the IOT Wearable Innovation Taxonomy is too

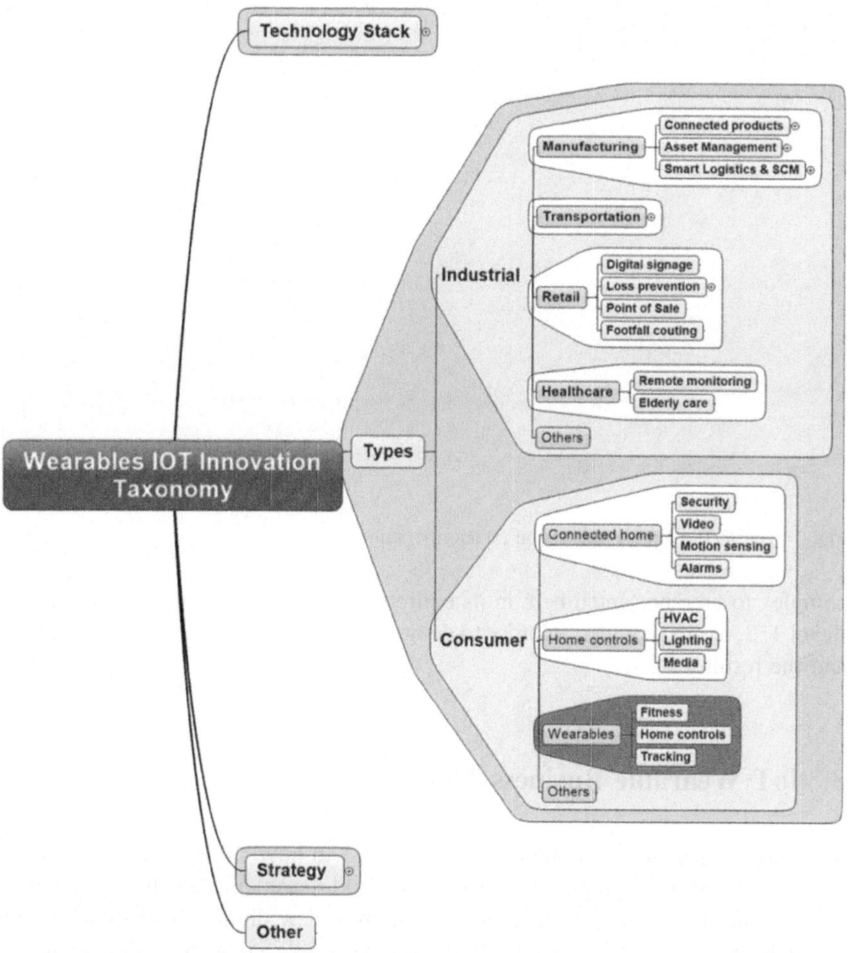

Fig. 2 The IOT wearable innovation taxonomy (types)

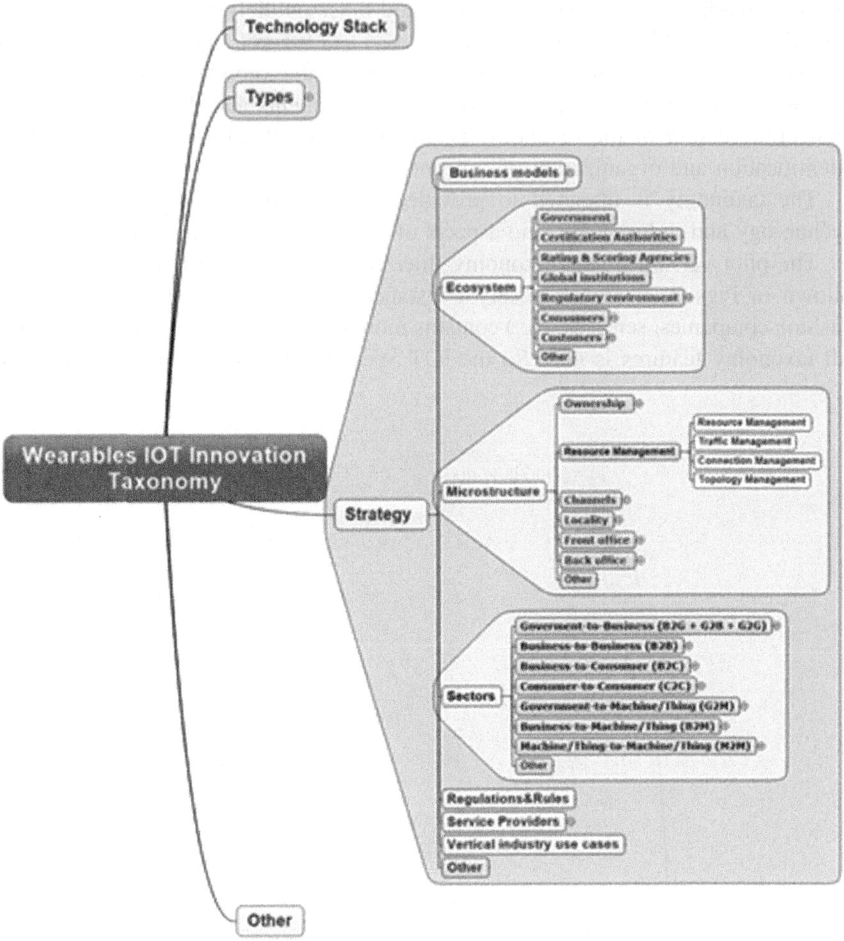

Fig. 3 The IOT wearable innovation taxonomy (strategy)

complex to be represented here in its entirety, but examples of the taxonomy parts (level 1, 2, 3) are provided in order to demonstrate both the process of classification and the result.

3 IoT Wearable Business Models

IoT raises a new set of strategic choices related to how value is created and captured, how the prodigious amount of new (and sensitive) data they generate is utilized and managed, how relationships through new digital channels with traditional business partners are redefined, and what role companies should play as industry boundaries are expanded (Porter and Heppelmann 2014; Hui 2014). IT and connectivity are becoming an integral part of the wearable product itself (Porter and Heppelmann 2014).

The business-model concept has been intensively developed since the beginning of the Internet as a conceptual tool or strategic framework (Timmers 1998, Teece 2010; Yablonsky 2014, 2016a, b). Osterwalder (2004) developed a business-model ontology, which is a conceptualization and formalization of the elements, relationships, vocabulary, and semantics of a digital business model, which is structured into several levels of decomposition with increasing depth and complexity. For now, business models have become mainstream and are connected to more generic research approaches and design perspectives (Bouwman et al. 2012) and are developed for IoT as well (Fleisch 2010; Fleisch et al. 2014; Chan 2015; Bilgeri et al. 2015; Hussain et al. 2015). To shorten the complicated ontology, Osterwalder and Pigneur (2010) have developed a simplified model of nine main building blocks named the Business Model Canvas (BMC). We add two new building blocks WHEN and WHERE to this model and represent it as a taxonomy (Fig. 4).

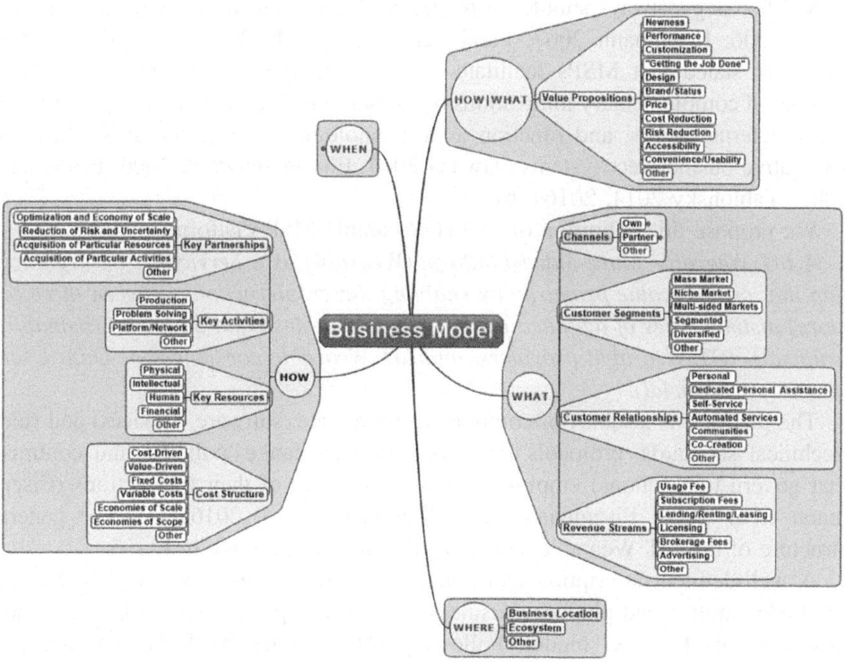

Fig. 4 Business-model building blocks (adopted from Osterwalder and Pigneur (2010), plus WHEN and WHERE building blocks)

Internet intermediaries or e-intermediaries can be defined as organizations bringing together or facilitating transactions between third parties on the internet (Yablonsky 2016b). According to OECD (OECD 2010), and cited by UNESCO (MacKinnon et al. 2015) "internet intermediaries bring together or facilitate transactions between third parties on the Internet. They give access to, host, transmit and index content, products and services originated by third parties on the Internet, or provide Internet-based services to third parties. They offer access to a host of activities through both wired and wireless technologies".

Currently in the e-business, market-intermediary stream, the dominant definition of an intermediary business models refers to platform intermediaries, where two or more agents interact through an intermediary two or multi-sided platform that forms two-sided and multi-sided markets (Armstrong 2006; Yablonsky 2016b). Hagiu and Wright (2015) proposed this definition of Multi-sided Platform (henceforth, MSP):

Multi-sided Platform is an organization that creates value primarily by enabling direct interactions between two (or more) distinct types of affiliated customers.

MSPs have two key attributes (Hagiu and Wright 2015; Yablonsky 2016a, b): (1) they enable direct interactions between two or more distinct sides, and (2) each side is affiliated with the platform.

MSP strategy always should be related with MSP business model (Eisenmann et al. 2006; Eisenmann 2007; Parker et al. 2016a, b; Yablonsky 2016a). It is explicitly stated that MSPs facilitate the generation of a potentially very large number of complementary innovations by tapping into the innovative capabilities of many external actors, and function as a technological foundation at the heart of innovative business ecosystems (Tiwana 2014; Eisenmann et al. 2006; Eisenmann 2007; Yablonsky 2014, 2016a, b).

We propose this definition of an IoT Wearable MSP Platform:

A IoT Wearable Multi-Sided Platform (Wearable as a Service) is an organization that creates value primarily by enabling direct interactions between three (or more) distinct types of affiliated actors: IoT Wearables Data Suppliers (human or artificial), IOT Wearable providers, and IoT Wearable consumers or Data Users (human or artificial).

The platform is a subset of components (hardware, software, services) and rules (technical standards, protocols for information exchange, policies, and contracts that govern transactions) employed by users in most of their transactions (Eisenmann et al. 2006; Eisenmann 2007; Yablonsky 2014, 2016a, b). The general structure of the IoT Wearable Three-sided Platform is shown in Fig. 5.

A well-defined description and conceptual information structuring of the IoT wearable, multi-sided-platform-business models are prerequisites for a common usage of this business model. Following (Osterwalder 2004; Osterwalder and Pigneur 2010; Gassmann et al. 2015; Fleisch et al. 2014; Dijkmana et al. 2015; Choudary 2015), we provide an IoT Wearable-MSP Business- Model Pattern (BPM), which is a conceptualization and formalization of the elements, relationships, vocabulary, and semantics of a possible IoT Wearable MSPs for the IoT domain (Fig. 6). We agree with Dijkmana et al. (2015) that the IoT boosts many

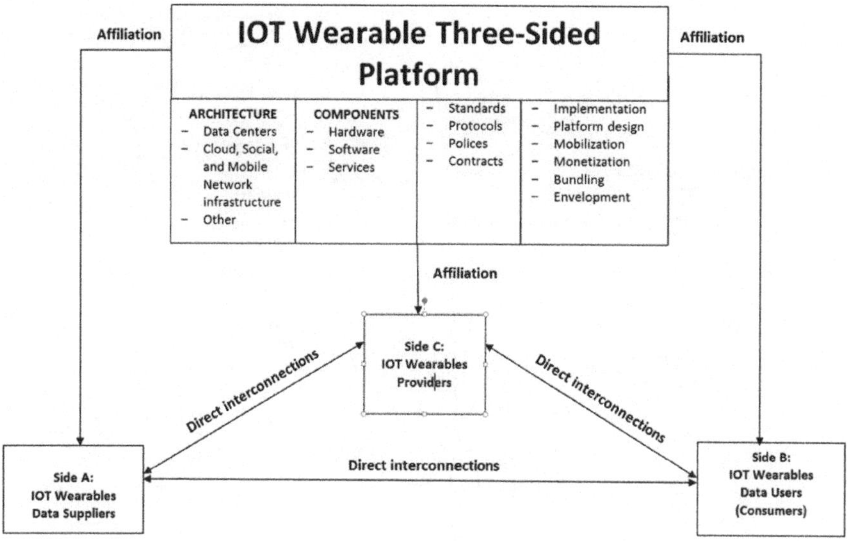

Fig. 5 IoT wearable three-sided platform

other "classic" business model patterns. Up to this point, we have described components and patterns that are newly possible with the IoT Wearables in the Table 1. But the main pattern for us is MSP BMP.

IoT-Wearable MSP BMP "classically" creates *value* by facilitating interactions between different groups/sides of the platform. Similar to known MSPs (Hagiu 2006; Osterwalder and Pigneur 2010; Evans 2013; Hagiu and Wright 2015; Yablonsky 2014, 2016a, b), the IoT Wearable MSP BMP *Value Preposition* creates value in such main areas:

- Mobilizing user groups-sides (*Customer Segments*) and balancing interests;
- Matchmaking between Customer Segments;
- Reducing costs by channeling transaction through the platform; and
- Strategic decisions in platform evolution, multi-homing costs, scaling, and liquidity.

IoT-Wearable MSP BMP has three or more Platform Customer Segments, each of which has its own Value Preposition and associated Revenue Stream. Furthermore, in some business models, one Customer Segment cannot exist without another. Sometimes one or more Customer Segments may enjoy free offers or reduced prices subsidized by revenues from other Customer Segments (see Table 1). Choosing which segment to subsidize can be a crucial, strategic, pricing decision that determines the success of the IoT Wearable MSP BMP. IoT-Wearable MSPs are of value to one group of customers only if the other groups of customers are also present.

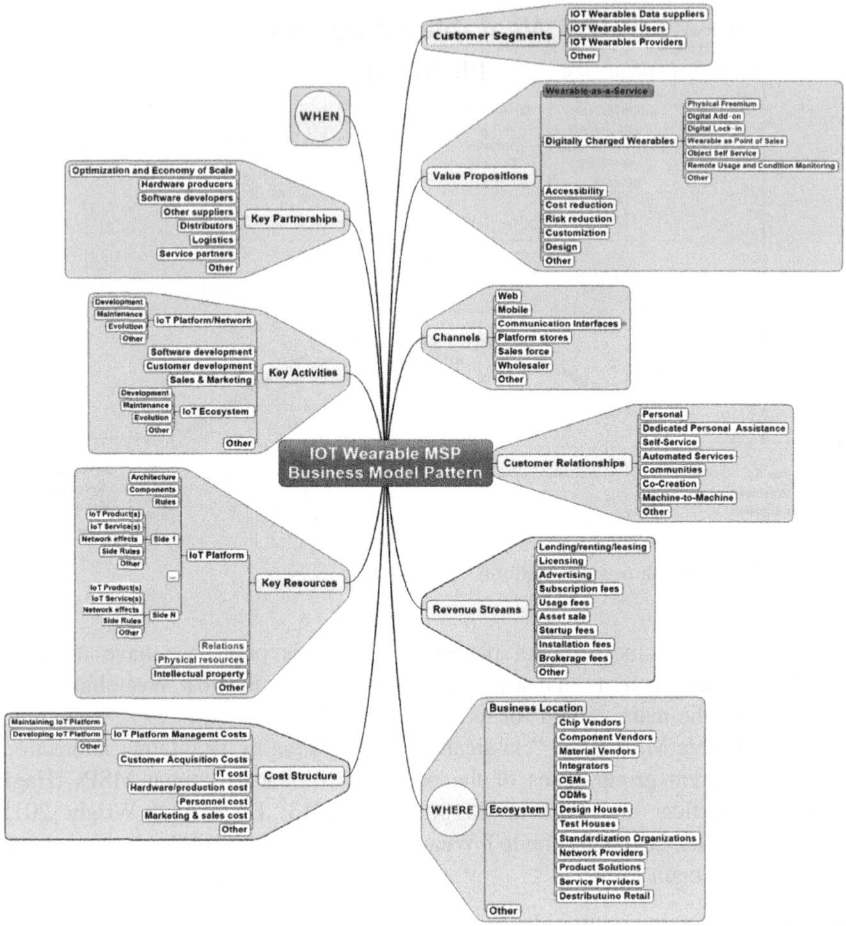

Fig. 6 IoT-wearable MSP BPN

The IoT Wearable MSP grows value to the extent that it attracts more users. IoTWearable MSP's value to a user depends on the number of other platform users and defines Willingness-to-Pay (henceforth, WTP) for platform participation. WTP for platform affiliation is a cap on platform fees (Eisenmann et al. 2006).

A network effect (also called network externality) is the effect that one user of a platform good or service has on the value of that product to other people (Eisenmann et al. 2006; Eisenmann 2007; Yablonsky 2014, 2016a, b). The key to IoT Wearable MSPs is to understand the network effects that occur when the value of a product depends on the number of other users (Shapiro and Varian 1999). IoT Wearable MSPs add personalization and context to the network effects between wearable products.

Table 1 Possible IoT wearables BPM types (adopted from Osterwalder 2004; Osterwalder and Pigneur 2010; Gassmann et al. 2015; Fleisch et al. 2014; Dijkmana et al. 2015)

Business model patterns	Facilitating components and patterns of the IoT wearables
Multi-sided Platform	"Wearable-as-a-Service"—Platforms combine data suppliers with data users and other possible actors
Crowdsourcing	"Wearable-as-a-Service"—A "crowd" of wearables generates data that is monetarized. Another type of MSP
Add-on	"Digital Add-on"—Remote sale and installation of additional options for wearables during the post-sale/usage period
Subscription	"Digital Add-on"—The usability of a wearable or sub-functions can be restricted to the time span of a subscription
Freemium	"Digital Add-on"—This business model can be applied in the physical world as well by combining free digital services with a physical wearable for sale. Premium services are available for a fee
Self-service	"Wearable Self-Service"—Wearable order consumables or services autonomously
Affiliation	"Wearable as Point of Sales"—Sales commissions for internet transactions are connected to the real world, e.g., the location of the user or a IoT wearable
Hidden revenue	"Wearable as Point of Sales"—For example, flexible, location-specific advertising becomes possible using the IoT technology
Customer loyalty	"Wearable as Point of Sales"—Customer loyalty can be rewarded not only for the purchase of a certain wearable but can be measured according to use of the wearable, or presence at a certain location, for instance
Fractionalized ownership	"Remote Usage and Condition Monitoring"—Use and consumption of wearables of lower value can be measured, making this business model applicable to those wearable as well
Guaranteed availability	"Remote Usage and Condition Monitoring"—Monitoring the status of production plants or equipment via the internet simplifies the application of the business-model pattern
Pay per use	"Remote Usage and Condition Monitoring"—Use and consumption of lower-value wearables can be measured, too. The business model pattern is applicable to these wearables as well
Performance-based contracting	"Remote Usage and Condition Monitoring"—Use and consumption of lower-value wearables can be measured also. Technology for monitoring the status of production plants and equipment further promotes the application of this business model pattern
Leverage customer data	"Wearable-as-a-Service"—Wearables transmit data to the manufacturer over their lifetime. The manufacturer can then use the data to improve the wearable
Lock-in	"Digital Lock-in"—Compatibility with competitors' systems is prevented by use of a digital handshake and authentication mechanism
Razor and blade	"Digital Lock-in"—"Razor blades" can be authenticated online using wearable digital mechanisms. Eliminates elaborate safeguarding of the business model, such as through patents, for instance

The main IoT Wearable MSP BMP *Costs* incurred under MSP BMP relate to maintaining and developing the platform and are characterized (Yablonsky 2016a, b) by:

- Homing costs, which are related to the adoption, operation, and/or other costs encountered due to platform affiliation (Armstrong 2006);
- Switching costs or the costs that consumers pay for switching from one platform to another (Shapiro and Varian 1999).

Various patterns including IOT Wearable MSP BMP make the free offer possible. Non-paying customers are financed by another part of the BM or by another Customer Segment of the IoT Wearable MSP BMP.

As it is known from the MSPs theory, the IoT Wearable MSP business model always should be related with MSP strategy (Jullien 2005; Eisenmann et al. 2006; Boudreau and Hagiu 2008; Eisenmann et al. 2011; Yablonsky 2014, 2016a, b) which consists of:

- Implementation
- Platform design
- Mobilization
- Monetization
- Bundling
- Envelopment.

As well as for any MSP (Yablonsky 2014, 2016a, b), there are several ways for IoT-Wearable MSPs to overcome entry barriers, such as strong network effects or high switching costs in multi-sided markets, namely:

- To provide technological features that are far superior to those of the dominant incumbent (Evans and Schmalensee 2002).
- To propose platform envelopment as another path of entry (Eisenmann et al. 2011). Envelopment happens when a new platform powers common elements (such as similar product functionalities or overlapping user bases) between different markets, through bundling or co-branding, in order to enter an adjoining market.

IoT Wearable MSPs enable new wearable products/services due to the reuse of platform components. They have lower fixed costs and enable shorter time to market for service providers.

Contemporary IoT Wearable MSP business models will have high evolutionary dynamics: the number of sides, products, and/or services is steadily increasing (Yablonsky 2014, 2016a, b). A successful IoT Wearable MSP should create some network of relationships among economic entities (producers, stakeholders, distributors, consumers, government agencies, etc.) that, through competition and/ or cooperation, facilitates the creation and distribution of a platform product or service (Yablonsky 2014, 2016a, b). The platform digital- ecosystem metaphor (Weill and Woerner 2015) could be used to describe such an IoT–Wearable-MSP network, which should be characterized by open, flexible, demand-driven, interactive, networked architecture, and collaborative environments (Tiwana 2014; Yablonsky 2014, 2016a, b). For some IoT Wearable MSPs, creation of the digital ecosystem would be a critical success factor for positive platform dynamics.

4 Main Findings and Recommendations

This paper provides one of the first ontologies/taxonomies of the IoT Wearable Innovation Taxonomy—a multidimensional framework, with a particular emphasis on technology stack, business model, product, service, and platform innovations. Moreover, it presents a framework that enables analysis of IoT Wearable innovations, which is also new to the academic world, and which may serve as a starting-point for further scholarly development. The goal of an ontology/taxonomy is not to insist on uniformity, but rather to know and model knowledge across the research area. In our analysis, we argue that rather than searching for a single acceptable definition of IoT wearables, a better approach would be to develop a classification system or taxonomy. A clear and precise description and structuring of the information in the IoT wearables domain are prerequisites for common research. Taxonomies and other types of controlled vocabularies are the preferred means to achieve such a common understanding by specifying the terms of the domain, disambiguating them from each other, controlling synonyms, and structuring the domain via term relationships. For conceptual grounding of the categories in the IoT Wearable, Innovation Taxonomy we use definitions from different information resources. Taxonomy includes a set of basic concepts, a set of relationships holding between those concepts, and a set of instances–international and local IoT-wearable- platform service providers. We suggest that, although classification systems have been used in the business and management disciplines, the more advanced quantitative methodologies have not yet been widely used. The future research could usefully build on these techniques to construct enhanced classification systems of IoT-wearable approaches across a variety of dimensions, in addition to the basic concepts of technology stack, business model, product, service, marketing, and platform innovations.

Using the IoT Wearable MSP BMP as the basis of the IoT Wearable business-models analysis, for each industry from the IoT Wearable Innovation Taxonomy, we update this pattern, converting it into an IoT Wearable MSP BMP for a specific type of industry platform (healthcare IoT Wearable MSP BMP, retail IoT Wearable MSP BMP, etc.).

Finally, for a given entity-a given industry IoT-Wearable MSP company- the industry MSP BMP is transferred to a company's MSP BM:

IoT Wearable MSP BMP— > industry IoT Wearable MSP BMP— > company IoTWearable MSP BM.

5 Limitations and Further Developments

We think that the early stage of the IoT Wearables' maturity limits our understanding of future developments in this area. From our point of view, the Standards, Protocols and Rules block in IoT Wearable MSP framework (see Fig. 5) will play a key role in the future IoT-enabled innovations for wearable product/service research.

6 Further Details

Research has been conducted with financial support from St. Petersburg State University grant (Project No. 16.38.378.2015).

References

451 Research (2015) 451 research explaining the internet of things ecosystem and taxonomy. https://451research.com/images/Marketing/IoT/IoT_Taxonomy_12.1.15.pdf. Accessed 25 Mar 2016

Armstrong M (2006) Competition in two-sided markets. RAND J Econ 37(3)

Bilgeri D, Brandt V, Lang M, Tesch J, Weinberger M (2015) The IoT business model builder. A White Paper of the Bosch IoT Lab in collaboration with Bosch Software Innovations GmbH

Boudreau K, Hagiu A (2008) Platform rules: multi-sided platforms as regulators. Available at SSRN 1269966

Bouwman H, De Reuver M, Solaimani S, Daas D, Haaker T, Novay, Janssen W, Iske P, Walenkamp (2012) Business models tooling and a research Agenda. In: 25th Bled eConference Special Issue, Bled, Slovenia

Bosch (2015) https://www.bosch-si.com. Accessed 25 Mar 2016

Chana M, Estèvea D, Fourniolsa J, Escribaa C, Campoa E (2012) Smart wearable systems: current status and future challenges. Artif Intell Med 56

Choudary S (2015) Platform scale: how an emerging business model helps startups build large empires with minimum investment. Platform Thinking Labs

Cisco (2015) http://www.cisco.com/c/m/ru_ru/internet-of-everything-ioe/. Accessed 25 Jan 2016

Díaz M, Martín C, Rubio B (2016) State-of-the-art, challenges, and open issues in the integration of Internet of things and cloud computing. J Netw Comput Appl. 25 Jan 2016

Dijkmana R, Sprenkelsa B, Peetersa T, Janssen A (2015) Business models for the internet of things. Int J Inf Manage 35:672–678

Eisenmann T (2007) Platform-mediated networks: definitions and core concepts. Harvard Business School, HBS: 807–049

Eisenmann T, Parker G, Van Alstyne M (2011) Platform envelopment. Strateg Manag J 32(12)

Eisenmann T, Parker G, Van Alstyne M (2006) Strategies for two-sided markets. Harvard Bus Rev 84(10):92–101

Evans DS, Schmalensee R (2002) Some economic aspects of antitrust analysis in dynamically competitive industries. In: Innovation policy and the economy, vol 2. MIT Press

Evans DS (2013) Economics of vertical restraints for multi-sided platforms. University of Chicago Institute for Law & Economics Olin, Research Paper No. 626

Fleisch (2010) What is the internet of things? An economic perspective, Auto-ID Labs White Paper WPBIZAPP-053, ETZ Zürich & University of St. Gallen, January 2010. http://www.im.ethz.ch/education/HS10/AUTOIDLABS-WP-BIZAPP-53.pdf

Fleisch E, Weinberger M, Wortmann F (2014) Business models and the internet of things, Bosch IoT Lab Whitepaper. http://www.IoT-lab.ch/?page_id=10543). Accessed 25 Mar 2016

Chan H (2015) Internet of things business models. J Serv Sci Manage 8

Gartner (2014) Gartner says the internet of things will transform the data center. http://www.gartner.com/newsroom/id/2684616. Accessed 25 Mar 2016

Gartner (2015) Research guide: the top 10 strategic technology trends for 2015. http://www.gartner.com/doc/2966917?refval=&pcp=mpe. Accessed 25 Mar 2016

Gartner (2016a) Building the digital platform: the 2016 CIO Agenda, Gartner 2016 CIO agenda survey, http://www.gartner.com/technology/cio-trends/cio-agenda/. Accessed 25 Mar 2016

Gartner, Jones N (2016b) Technologies, architectures & vendors for the internet of things. Gartner webinar

Gartner, Cearley D (2016c) The gartner top 10 strategic technology trends for 2016. Gartner webinar

GE (2015) GE 2015. Annual Report

Google (2015) https://cloud.google.com/solutions/IoT/. Accessed 25 Mar 2016

Gassmann O, Frankenberger KK, Csik MM (2015) The business model navigator: 55 models that will revolutionise your business. FT Press, 400 p

Green J, McCarson B, Devine M (2014) Building the internet of things. https://daue6ehqissah.cloudfront.net/breakouts/2014/H-ARC-01_Cisco-Intel-IBM_FINAL.pdf. Accessed 25 Mar 2016

Hagiu A (2006) Pricing and commitment by two-sided platforms. RAND J Econ 37(3)

Hagiu A, Wright J (2015) Multi-sided platforms. Int J Ind Org

Hao Y, Helo P (2015, in press) The role of wearable devices in meeting the needs of cloud manufacturing: a case study. Rob Comput Integr Manuf.Accessed 29 Oct 2015

Hui G (2014) How the internet of things changes business models. Harv Bus Rev

Hussain M, Al-Haiqi F, Zaidan A, Zaidan B, Kiah V, Anuar N, Abdulnabi M (2015) The land-scape of research on smartphone medical apps: Coherent taxonomy, motivations, open challenges and recommendations. Comput Methods Programs Biomed 122(3):393–408

IBM (2015) IBM Watson internet of things (IoT). http://www.ibm.com/internet-of-things/. Accessed 29 Oct 2015

IDC (2015) IDC's worldwide internet of things taxonomy. https://www.idc.com/getdoc.jsp?containerId=256186. Accessed 25 Mar 2016

IEEE (2015) IEEE-SA internet of things (IoT) ecosystem study'. http://standards.ieee.org/innovate/IoT/. Accessed 25 Mar 2016

Jullien B (2005) Two-sided markets and electronic intermediaries. CESifo Econ Stud 51(2–3):233–260

Lee I, Lee K (2015) The Internet of Things (IoT): Applications, investments, and challenges for enterprises. Bus Horiz 58(4):431–440

Industrie (2013) Recommendations for implementing the strategic initiative INDUSTRIE 4.0, Final report of the Industrie 4.0 Working Group, http://www.acatech.de/fileadmin/user_upload/Baumstruktur_nach_Website/Acatech/root/de/Material_fuer_Sonderseiten/Industrie_4.0/Final_report/Industrie_4.0_accessible.pdf. Accessed 25 Mar 2016

Experton (2016) Experton industry 4.0/internet of things vendor benchmark 2016. https://www.bosch-si.com/media/en/bosch_software_innovations/documents/white_paper/manufacturing_3/experton_group_industry_4_0_IoT_benchmark_2016.pdf. Accessed 25 Mar 2016

Puustjärvi J, Puustjärvi L (2015) The role of smart data in smart home: health monitoring case. Proc Comput Sci 69:143–151

LexInnova (2015) Wearable technology. http://www.lex-innova.com/resources-reports/. Accessed 25 Mar 2016

MacGillivray C, Turner V, Kraus B, Kumar M (2015) IDC's worldwide internet of things taxonomy, 12 p

MacKinnon R, Hickok E, Bar A, Lim H (2015) Fostering freedom online. The role of internet intermediaries, A report prepared for UNESCO's Division for Freedom of Expression and Media Development. http://unesdoc.unesco.org/images/0023/002311/231162e.pdf. Accessed 25 Mar 2016

Microsoft (2015) https://azure.microsoft.com/en-us/services/IoT-hub/. Accessed 25 Mar 2016. Miorandia D, Sicarib S, De Pellegrinia F, Chlamtaca I (2015) Internet of things: vision, applications and research challenges. Ad Hoc Netw 33(10):87–111

OECD (2010) The economic and social role of internet intermediaries (Paper No. 171). OECD digital economy

Osterwalder A (2004) The business model ontology—a proposition in a design science approach, PhD thesis. University of Lausanne, Ecole des Hautes Etudes Commerciales HEC

Osterwalder A, Pigneur I (2010) Business model generation: a handbook for visionaries, game changers, and challengers. Wiley

Parker G, Van Alstyne M, Choudary S (2016a) Platform revolution: how networked markets are transforming the economy and how to make them work for you. W. W. Norton & Company

Parker G, Van Alstyne M, Choudary S (2016b) Pipelines, platforms, and the new rules of strategy. Harv Bus Rev 94(4)

Peitz M, Waldfogel J (2012) The Oxford handbook of the digital economy. Oxford University. Poli R, Healy M, Kameas A (2010) Theory and applications of ontology: computer applications set: theory and applications of ontology, vol XVIII, 576 p. Springer. Porter M (2001) Strategy and the internet. Harv Bus Rev 79(3)

Porter M, Heppelmann J (2014) How smart, connected products are transforming competition. Harv Bus Rev 92(11)

Shapiro C, Varian H (1999) Information Rules. Harvard Business School Press, Boston

Stock T, Seliger G (2016) Opportunities of sustainable manufacturing in industry 4.0, Procedia CIRP, Volume 40, 2016, pp 536–541

Taxonomy (2016) https://cantina.co/the-internet-of-things-our-take-on-a-taxonomy/. Accessed 25 Mar 2016

Timmers P (1998) Business models for electronic markets. Electron Mark 8(2):3–8. Tiwana A (2014) Platform ecosystems: aligning architecture, governance, and strategy. Morgan Kaufmann Publishers

Teece D (2010) Business models, business strategy and innovation. Long Range Plan 43(2)

Ugur S (2013) Wearing embodied emotions: a practice based design research on wearable technology. Springer, 112 p

Weill P, Woerner S (2015) Thriving in an increasingly digital ecosystem. MIT Sloan Manage Rev 56(4)

Y.2060 (2012) Y.2060: overview of the internet of things, recommendation Y.2060 (06/12). https://www.itu.int/rec/T-REC-Y.2060-201206-I. Accessed 25 Mar 2016

Yablonsky S (2014) E-finance innovation services in Russia. Int J Bus Innov Res (IJBIR) 8(5)

Yablonsky S (2016a) Multi-sided search platforms: global and local. Int J Technol Mark (IJTMKT) 11(3)

Yablonsky S (2016b) Intermediaries in E-Commerce, In: In Lee (ed.) Encyclopedia of E-Commerce Development, Implementation, and Management, Volume I, IGI Global

Part III
Fashion Operations and Supply Chain Management

Linking Inventory Management Performance and Operational Performance: An Empirical Analysis of U.S. Fashion Apparel and Accessory Industries

Fethi Çalışır and Gülşah Hançerlioğulları

Abstract Managing inventories is at the core of operational performance in fashion industries. Due to its importance in practice, inventory management has been a well-studied area of research in operations management. The purpose of this study is to examine the relationship between inventory management performance including inventory efficiency, productivity and responsive, and firm operational performance. We present and empirically test a performance model which integrates the various dimensions of a fashion industry's inventory management execution. The regression analysis is used to study the effect of various measures on inventory performance. We use financial data for 40 publicly listed U.S. fashion apparel and accessory industries for the 6-year period, 2010–2015, from "Compustat North America Annually Updated" available at Standard and Poor's Compustat database using Wharton Research Data Services (WRDS). We discuss the implications of these empirical results on the study of inventory policy execution, and propose some guidance for further research.

Keywords Inventory management · Fashion apparel industry · Operations management · Operational performance · Inventory efficiency · Inventory productivity · Inventory responsiveness

F. Çalışır · G. Hançerlioğulları (✉)
Department of Industrial Engineering, Faculty of Management, Istanbul Technical University, 34367 Istanbul, Turkey
e-mail: ghancerliogullari@itu.edu.tr

F. Çalışır
e-mail: calisirfet@itu.edu.tr

R. Rinaldi and R. Bandinelli (eds.), *Business Models and ICT Technologies for the Fashion Supply Chain*, Lecture Notes in Electrical Engineering 413,
DOI 10.1007/978-3-319-48511-9_13

1 Introduction

A longitudinal approach to examining inventory management performance is important to understand fashion apparel and accessory firm competitive performance. In this paper we explore the relationship between the inventory management of U.S. fashion apparel and accessory firms and competitive operational performance advantages. Some research has identified a number of firm-level inventory management issues in retailing, such as industry competitiveness, gross margin, capital investment intensity and sales surprise (Fisher and Raman 2001; Olivares and Cachon 2009; Gaur et al. 1999, 2005; Eroglu and Hofer 2011; Rajagopalan 2012; Hancerliogullari et al. 2016). The objective of this paper to show and empirically test a comprehensive performance model that incorporates the different dimensions of a fashion firm's inventory management execution including efficiency, productivity and responsiveness in order to evaluate the inventory management performance effect on firm competitive outperformance. By applying our model to fashion apparel and accessory industry, we expect to build on previous literature supporting a more scientific understanding of firm performance benchmarking and evaluation in this distinctive sector (Fisher and Raman 2010).

2 Data Description and Definition of Variables

We obtained the financial data for all publicly listed U.S. fashion apparel and accessory industries for the 6-year period 2010–2015 from "Compustat North America Annually Updated" available at Standard & Poor's Compustat database using Wharton Research Data Services (WRDS). The U.S. Department of Commerce assigns a Standard Industry Classification (SIC) code to each firm according to its primary industry segment. We group together firms in similar product groups as there are substantial overlaps among their products. For example, all firms with SIC codes between 5600–5699 are collected in a segment called "Fashion apparel and accessory industries". The categorization that we use is the similar as that was used in Gaur et al. (2005). Table 1 summarizes the segment, corresponding SIC codes, and a few examples of firms in each category. The original data set contained

Table 1 Classification of the U.S. fashion apparel and accessory industries

Industry name	SIC codes	Examples of firms
Apparel and accessory stores	5600	Claire's, burlington stores, american eagle outfitters
Womens' clothing stores	5621	Charming shoppes, coldwater creek, New York & Co
Family clothing stores	5651	Gap, nordstrom, ross stores, TJX companies
Shoe stores	5661	Foot locker, DSW, finish line

290 annual observations across 57 firms. We omit from our data set the firms that have less than five consecutive years of data. Our final data set contains 234 annual observations across 40 firms for the period 2010–2015.

2.1 *Model Variables*

Operations Management literature proposes that inventory efficiency, gross margin productivity and responsiveness are positively correlated with firm's operational performance. However, it is unclear how to link this to fashion industries' firm productivity performance. By applying the model provided in Shockley and Turner (2015) to U.S. fashion apparel and accessory industries, we hope to build on existing literature. For purposes of our study, we used the following model variables; their COMPUSTAT code and definitions are provided in Table 2.

2.2 *Inventory Policy Performance Variables*

We use three metrics as proxies to measure the different inventory performance dimensions, inventory policy performance variables and their calculations are provided in Table 3.

Table 2 Definition of model variables

Variables	COMPUSTAT code	Definition
$\mathrm{Inv}_{\mathrm{it}}$	INVT	Inventories-total
S_{it}	SALE	Sales/turnover (net)
$\mathrm{COGS}_{\mathrm{it}}$	COGS	Cost of goods sold
$\mathrm{AT}_{\mathrm{it}}$	AT	Assets-total
$\mathrm{LIFO}_{\mathrm{it}}$	LIFR	LIFO reserve
$\mathrm{EBITDA}_{\mathrm{it}}$	EBITDA	Earnings before interest
$\mathrm{PPE}_{\mathrm{it}}$	PPENT	Property, plant and equipment-total (net)

Table 3 Definition of inventory policy performance variables

Inventory measures	Calculations
Inventory efficiency (inventory/COGS ratio)	$XI_{it} = \frac{Inv_{it}}{COGS_{it}}$
Gross margin return on inventory investment	$GMROII_{it} = \frac{GM_{it}}{Inv_{it}}$
Inventory responsiveness (co-movements of inventory and COGS)	$XC_{it} = \frac{\mathrm{Inv}_{\mathrm{it}} - \mathrm{Inv}_{\mathrm{i(t-1)}}}{\mathrm{Inv}_{\mathrm{i(t-1)}}} - \frac{\mathrm{COGS}_{\mathrm{it}} - \mathrm{COGS}_{\mathrm{i(t-1)}}}{\mathrm{COGS}_{\mathrm{i(t-1)}}}$
Inventory responsiveness (over-responsiveness)	$XC+_{it} = XC_{it} \times 1_{(IR \geq 0)}$
Inventory responsiveness (under-responsiveness)	$XC-_{it} = XC_{it} \times -1_{(IR \geq 0)}$

Relative inventory level (XI): Measures the total average inventory over cost of goods sold and is the inverse of inventory turnover. It is a common performance indicator reflecting inventory efficiency and leanness.

Gross margin return on inventory investment (GMROII): Measures how much profit contribution a firm earns on every dollar it spends on inventory. It evaluates the profit-productivity of the inventory sold.

Inventory responsiveness (XC): Measures the responsiveness of the firm in matching customer sales with the inventory held by the firm over the annual period. Two measures (XC^+, XC^-) specify how quickly a firm adjusts inventory levels in response to annual changes in the sales environment.

We made several transformations to conduct our analysis including that every balance sheet item was adjusted to get an average value for that item for each firm for the annual period. For instance, the inventory measure used in each variable was calculated based on averaging the prior year period-ending inventory ($Inv_{i(t-1)}$) and the current year (Inv_{it}) period-ending inventory balances (adjusted for the LIFO reserve). Similarly, same procedure was done for the cost of goods sold and gross margin are adjusted for the LIFO reserve as stated. Table 4 shows the descriptive statistics for each industry for the performance variables. We see that there is a great level of variation in inventory (minimum average at 205,185 in womens' clothing and maximum average at 609,715 in family clothing) across different segments of the fashion industry.

2.3 Firm and Segment Control Variables

In order to manage total inventories, total capital investment and sales growth for each firm are also controlled. Existing research has stated the importance of capital investment variable when evaluating inventory and operational performance

Table 4 Summary statistics of the inventory performance variables

Industry	SIC codes	# of firms	# of observations	Avg. annual sales ($ mil.)	Avg. inventory	Avg. XI	Avg. GMROII	Avg. XC
Apparel and accessory stores	5600	6	35	1772,552	254,317	0,232	0,003	0,036
Womens' clothing stores	5621	13	75	2035,592	205,185	0,191	0,008	0,014
Family clothing stores	5651	16	94	4891,776	609,715	0,294	0,005	0,031
Shoe stores	5661	5	30	2507,658	489,861	0,338	0,001	0,001
All		40	234	3204,126	411,534	0,257	0,005	0,022

Table 5 Definition of control variables

Control variables	Calculations
Capital investment (firm operational capital)	$K_{it} = log\left[PPE_{it} + \sum_{\tau=1}^{5} \frac{LC_{it}}{1+r^{\tau}}\right]$
Sales growth rate (firm revenue growth)	$\Delta S_{it} = [S_{it} - S_{i,t-1}]/S_{i,t-1}$
Relative gross margin (firm vs. segment avg.)	$GM\%_{it} = \frac{S_{it} - COGS_{it}}{S_{it}}$ $rGM_{it} = GM_{it} - GM_{seg,t}$
Relative sales over fixed assets (firm vs. segment avg.)	$SOA_{it} = S_{it}/(AT_{it} - Inv_{it})$ $rSOA_{it} = SOA_{it} - SOA_{seg,t}$

(Kesavan et al. 2010; Eroglu and Hofer 2011; Hancerliogullari et al. 2016). As being discussed in the literature, K_{it} is the firm's total capital investment, the log of the sum total of total property, plant and equipment and the net present value of five-year lease contracts (operating leases) using the notation $LC_{it,1}(MRC1), \ldots, LC_{it,5}(MRC5)$ in COMPUSTAT while the weighted average cost of capital of the fashion industry (r = 8.25%) reported from Value Line as the annual discount rate.

Firm's revenue sales growth should be correlated with greater operating performance in fashion industry; therefore, fashion analysts focus on store growth and store sales growth. ΔS_{it} is the revenue sales growth of the fashion company in period t from period t−1. Moreover, Gaur et al. (2005) and Kesavan et al. (2010) indicate that controlling the gross margin of the product portfolio and the non-inventory fixed-asset performance of the firms is important. Therefore, segment-adjusted gross margin, rGM_{it}, and segment-adjusted non-inventory fixed assets, $rSOA_{it}$, are used as additional control variables. These variables and their calculations are summarized in Table 5.

2.4 Dependent Variables

As dependent variables, we use both the firm's segment adjusted return on assets (rROA) and return on sales (rROS) to measure performance. ROA is defined as earnings before interest generated per dollar of total asset investment. On the other hand, ROS is the earnings before interest generated for every dollar in annual sales. The correlation matrix is listed in Table 6.

Table 6 Correlation matrix of key model variables

	rROA	rROS	XI	GMROII	XC^+	XC^-	**ΔS**	rGM	rSOA
rROA	1								
rROS	0.91*	1							
XI	−0.22*	−0.20*	1						
GMROII	0.13	−0.05	−0.05	1					
XC^+	−0.10	−0.06	0.27*	0.04	1				
XC^-	0.10	0.06	−0.27*	−0.04	−1*	1			
ΔS	0.59*	0.54*	−0.05	0.17*	−0.13	0.13*	1		
rGM	0.27*	0.31*	0.34*	0.16*	−0.06	0.06	0.32*	1	
rSOA	0.04	−0.23*	0.07	0.09	−0.01	0.01	−0.00	−0.22	1

*$p < 0.05$

3 Hypothesis Development and Model Specification

In order to demonstrate how individual fashion apparel and accessory firms manage inventory responsiveness, inventory gross margin productivity and inventory leanness in various ways for competitive advantage, we present three hypotheses, which were presented and tested earlier in other empirical papers including Shockley and Turner (2015). We provide how effective inventory management may contribute to superior firm-level operational performance advantage for a given period. The hypotheses that we develop in this section are mainly inspired by the mathematical models of inventory theory.

We state the hypotheses examining the relationship of inventory efficiency to operating performance in U.S. fashion industries as:

Hypothesis 1. *A firm's relative inventory (XI) measure is negatively correlated with firm outperformance.*
Hypothesis 2. *A firm's gross margin return on inventory investment measure is positively correlated with firm outperformance.*
Hypothesis 3. *A firm's relative over-responsiveness (XC^+) or under-responsiveness (XC^-) is indicative of worse inventory management responsiveness and is negatively correlated with firm outperformance.*

$$rROA_{its} = b^1 \; Independent\ variables_{it} + b^2 K_{it} + b^3 \Delta S_{it} + b^4 rGM_{it} + b^5 rSOA_{it} + \varepsilon_{it} \tag{1}$$

$$rROS_{its} = b^1 \; Independent\ variables_{it} + b^2 K_{it} + b^3 \Delta S_{it} + b^4 rGM_{it} + b^5 rSOA_{it} + \varepsilon_{it} \tag{2}$$

We develop linear regression models Eqs. (1)–(6), where dependent variables $rROA_{its}$ and $rROS_{its}$ are the firm's segment-adjusted ROA and ROS performance. b^1 is the coefficient for each of the firm-specific indicator variables for inventory management performance; in other words, XI, GMROII, XC; b^2 and b^3 denote the coefficients for total individual firm capital investment (K_{it}) and change in firm sales from the prior year (ΔS_{it}), respectively; lastly b^4 and b^5 are the coefficients for the segment-adjusted firm-level control variables, gross margin (rGM_{it}) and segment-adjusted non-inventory fixed assets $(rSOA_{it})$ respectively; and ε_{it} is random model error.

Model 1 (efficiency)

$$rROA_{its} = b^1 XI_{it} + b^2 K_{it} + b^3 \Delta S_{it} + b^4 rGM_{it} + b^5 rSOA_{it} + \varepsilon_{it} \quad (3)$$

$$rROS_{its} = b^1 XI_{it} + b^2 K_{it} + b^3 \Delta S_{it} + b^4 rGM_{it} + b^5 rSOA_{it} + \varepsilon_{it} \quad (4)$$

Model 2 (productivity)

$$rROA_{its} = b^1 GMROII_{it} + b^2 K_{it} + b^3 \Delta S_{it} + b^4 rGM_{it} + b^5 rSOA_{it} + \varepsilon_{it} \quad (5)$$

$$rROS_{its} = b^1 GMROII_{it} + b^2 K_{it} + b^3 \Delta S_{it} + b^4 rGM_{it} + b^5 rSOA_{it} + \varepsilon_{it} \quad (6)$$

Model 3 (responsiveness)

$$rROA_{its} = b^1 XC_{it} + b^2 K_{it} + b^3 \Delta S_{it} + b^4 rGM_{it} + b^5 rSOA_{it} + \varepsilon_{it} \quad (7)$$

$$rROS_{its} = b^1 XC_{it} + b^2 K_{it} + b^3 \Delta S_{it} + b^4 rGM_{it} + b^5 rSOA_{it} + \varepsilon_{it} \quad (8)$$

4 Results

Our study shows strong support for H1 and H2 as each inventory performance indicator is a statistically significant predictor of firm outperformance. The inventory efficiency relative inventory measure (XI) is significant ($p < 0.05$) and negatively correlated with operating performance. The inventory productivity measure, gross margin return on investment (GMROII) is significant ($p < 0.05$) and positively correlated with operating performance. These results are consistent when using rROA or rROS as a dependent variable. On the other hand, H3 is not supported, and is weak when using neither rROA nor rROS as a dependent variable. XC^+ does not indicate worse rROA or rROS performance. Overall, the results for inventory responsiveness suggests that fashion firms that over-corrected in their inventory relative to sales changes outperform competitors (Tables 7 and 8).

Table 7 Coefficient estimates for Models 1, 2 and 3 with relative ROA performance

	Model 1		Model 2		Model 3	
	Coefficient	t-value	Coefficient	t-value	Coefficient	t-value
XI	−0.285*	−4.86				
GMROII			2.974*	3.00		
XC^+					0.011	0.14
XC^-						
K	0.042*	2.61	0.094*	4.59	0.057*	3.39
ΔS	0.818*	9.33	0.837*	9.19	0.899*	9.87
rGM	0.306*	3.77	0.085	1.10	0.138	1.79
rSOA	0.021*	3.18	0.018*	2.69	0.016*	2.45
Constant	0.112*	2.19	−0.123*	−2.03	−0.006	−0.14
R^2	45.66%		42.31%		40.05%	
R^2 (adj)	44.47%		41.05%		38.73%	

*$p < 0.05$

Table 8 Coefficient estimates for models 1, 2 and 3 with relative ROS performance

	Model 1		Model 2		Model 3	
	Coefficient	t-value	Coefficient	t-value	Coefficient	t-value
XI	−0.092*	−3.47				
GMROII			0.608	1.36		
XC^+					0.043	1.14
XC^-						
K	0.030*	4.22	0.043	4.68*	0.037*	4.98
ΔS	0.337*	8.51	0.351	8.59*	0.368*	9.18
rGM	0.107*	2.91	0.041	1.20	0.053	1.56
rSOA	−0.005*	−1.99	−0.007	−2.31*	−0.007*	−2.31
Constant	0.018	0.78	−0.044	−1.61	−0.025	−1.18
R^2	45.02%		42.59%		42.44%	
R^2 (adj)	43.82%		41.33%		41.18%	

*$p < 0.05$

5 Discussion and Conclusion

In this paper, we have applied a model provided in Shockley and Turner (2015) to the U.S. fashion apparel and accessory industries in order to observe the relationship between inventory management and operational performance advantages. We have contributed to inventory management research by observing the impact of inventory management performance indicators over a 6-year period 2010–2015. Our result indicates that inventory efficiency and productivity significantly impact firm operating performance. In response to Hypothesis 1, it is found that firms that operate a leaner and more efficient inventory system significantly outperform competitors. Similarly, in response to Hypothesis 2, firms having higher gross margin returns on inventory investment obtain superior operating performance.

References

Eroglu C, Hofer C (2011) Inventory types and firm performance: vector autoregressive and vector error correction models. J Bus Logist 32(3):227–239

Fisher M, Raman A (2001) Introduction to focused issue: retail operations management. Manufact Serv Oper Manage 3(3):189–190

Fisher ML, Raman A (2010) The new science of retailing. Harvard Business Press. Boston, MA

Gaur V, Fisher M, Raman A (1999) What explains superior retail performance? Archive@NYU OM-2005-3:1-39

Gaur V, Fisher M, Raman A (2005) An econometric analysis of inventory turnover performance in retail services. Manage Sci 51(2):181–194

Hancerliogullari G, Sen A, Aktunc EA (2016) Demand uncertainty and inventory turnover performance: an empirical analysis of the U.S. retail industry. Int J Phys Distrib Logist Manage 46(6/7):681–708

Kesavan S, Gaur V, Raman A (2010) Do inventory and gross margin data improve sales forecasts for US public retailers? Manage Sci 56(9):1519–1533

Olivares M, Cachon G (2009) Competing retailers and inventory: an empirical investigation of General Motors' dealership in isolated U.S. markets. Manage Sci 55(9):1586–1604

Rajagopalan S (2012) Impact of variety and distribution system characteristics on inventory levels at U.S. Retailers. Manufact Serv Oper Manage 15(2):191–204

Shockley J, Turner T (2015) Linking inventory efficiency, productivity and responsiveness to retail firm outperformances: empirical insights from U.S. retailing segments. Prod Planning Control 26(5):393–406

Wharton Research Data Service (WRDS) (2009) University of Pennsylvania. https://wrds-web.wharton.upenn.edu/wrds. Accessed 22 Mar 2016

Logistics Solutions to Support Cross Border E-Commerce Towards China: The Case of the Apparel Industry

Maria Giuffrida, Riccardo Mangiaracina, Alessandro Perego and Angela Tumino

Abstract *Purpose*: In the past few years China has experienced a significant growth of both its economy and the adoption of technologies. With 310 billion dollars turnover it has become one of the biggest B2C ecommerce markets in the world (WorldPay 2015). As Chinese consumers are generally leaned towards buying international brands, entering this market while exploiting the rising trend of online and mobile commerce adoption might be an interesting opportunity for western fashion companies. However, the ultimate success of an e-commerce initiative is strictly dependent on the effectiveness of the distribution process. Geographical and cultural distance, expected service level, tariff and non-tariff barriers are some of the elements making this activity more complex in a global environment. The aim of this paper is to provide a quantitative model to compare different logistic solutions underlying a B2C e-commerce initiative in China. The analysis takes the perspective of a European firm operating in the apparel industry. *Research Approach*: The paper has an empirical approach. It presents an Activity-based model to quantify the logistics costs of three logistics solutions that can be adopted to sell online in an international context. The alternatives have been identified through interviews and context data have been derived from case studies. The model has been eventually validated by industry experts and practitioners. The considered logistics options are (i) distribution from a warehouse located in China, (ii) distribution through a sorting hub located in China, and (iii) distribution from a warehouse in the Country of origin through express couriers. By elaborating some specific data (e.g. weight and volume of the garment) as inputs, the model returns

M. Giuffrida (✉) · R. Mangiaracina · A. Perego · A. Tumino
Department of Management, Economics and Industrial Engineering, Politecnico di Milano, Via Raffaele Lambruschini, 4b, Milan, Italy
e-mail: maria.giuffrida@polimi.it

R. Mangiaracina
e-mail: riccardo.mangiaracina@polimi.it

A. Perego
e-mail: alessandro.perego@polimi.it

A. Tumino
e-mail: angela.tumino@polimi.it

R. Rinaldi and R. Bandinelli (eds.), *Business Models and ICT Technologies for the Fashion Supply Chain*, Lecture Notes in Electrical Engineering 413,
DOI 10.1007/978-3-319-48511-9_14

the distributive solution associated to the lowest logistics costs. In addition, a sensitivity analysis is provided to take into account variations in both the annual demand of the item and its value. *Originality*: This study simultaneously addresses three major subjects which are relevant for a fashion company, i.e. the design of a logistics strategy, the use of e-commerce and the implementation of an internationalisation project. A preliminary literature review has revealed that these themes have largely been debated as stand-alone topics while the "intersection" of these fields has not been adequately investigated yet. *Research Impact*: The present study contributes to create knowledge in the field of technology-driven internationalisation strategies. At the same time, it paves the way to additional meaningful research that can be originated by addressing some of its limitations. For instance, one of the hints for future research on this topic is to evaluate the impact of product returns which can be complex to manage at an international level. *Practical Impact*: The main beneficiaries of this study are producers and/or retailers operating in the apparel industry. They can learn some of the logistics alternatives to serve the promising Chinese B2C e-commerce market and identify the most efficient strategy for them.

Keywords Internationalisation strategy · B2C e-commerce · China · Apparel industry · Distribution network · Logistics · Quantitative model

1 Introduction

International trade is one of the key features of our increasingly globalised society and one of the main contributors to the world Gross Domestic Product (GDP). Data from the World Trade Organisation (WTO) indeed show that, despite the financial crisis, the ratio of global trade of goods and services to GDP has increased significantly in the past decade rising from 20% in 1995 to 30% in 2014 (WTO 2015).

Despite positive, this trend could possibly be faster if combined with another of the most significant recent phenomena, i.e. e-commerce. In 2014, the ratio of B2C e-commerce to the world GDP was around 2.64%, which is still a limited value but with significantly higher growth rates, i.e. more than 100% since 2010 (E-commerce Foundation 2015). This is a signal of the potential future developments of this technology, which should not be seen only as an opportunity for manufacturers and retailers to grow locally, but also to enter new markets (Premazzi et al. 2010). On a global scale, the development of the internet depends not only on its technological evolution, but also on the value generated for the different players involved (Ghezzi et al 2013). Nonetheless, e-commerce represents an interesting tool in support of internationalisation especially if we focus on certain countries and industries that seem more "digital oriented" than others.

China, for instance, is one of the countries commonly acknowledged as very promising for the adoption of e-commerce. In the past few years, it has experienced a significant growth of both its economy and the use of technologies. With 566 billion euro turnover, it has become one of the biggest B2C e-commerce markets in the world (Forrester 2015; eMarketer 2015). As Chinese consumers are generally leaned towards buying international brands, entering this market while exploiting the rising trend of online and mobile commerce adoption might be an interesting opportunity for western companies.

Players in the fashion industry can be listed among the greatest potential beneficiaries of Chinese online cross border shopping propensity. Indeed, purchases of fashion items, especially luxury ones, are very frequent in China, where the high-end fashion consumption is among the highest in the world (Bonetti 2014). Despite being one of the most active industries in the e-commerce landscape, fashion is not easy to sell online. E-commerce indeed does not eliminate the features of the fashion supply chain, commonly recognised as highly complex and characterised by demanding logistics requirements (Ghezzi et al. 2012). This holds even truer when operating on a global scale. The "problem" of physically shipping products overseas stays regardless of the adopted trade channel. Rather, the longer the distance (not only geographical) to the destination market, the higher the complexity. This is why an effective configuration of the logistics underlying a cross border online operation is key for the ultimate success of the initiative.

The aim of the present paper is to add knowledge to this topic by performing a quantitative comparison of possible logistics strategies supporting online export to China. This addresses the need to bridge some gaps identified in present literature.

Internationalisation through e-commerce is indeed an issue of growing importance, but has received limited attention in literature so far.

This paper is organised as follows: Sect. 2 specifies the research questions stemming from a preliminary literature review in the field of cross border B2C e-commerce logistics. Section 3 describes the research phases and the adopted methodology. Section 4 contains the description and discussion of results while Sect. 5 concludes.

2 Theoretical Background and Research Questions

A preliminary analysis of up-to-date literature reveals there are plenty of contributions investigating the link between logistics and e-commerce. There is an overall agreement on the fact that logistics is a crucial aspect of an e-commerce initiative (e.g. Cho et al. 2008, Ramanathan et al 2014). Its relevance is actually so high that it is often listed among the factors that might cause e-commerce failures, when not properly managed (Delfmann et al. 2002). Indeed logistics is considered a source of competitive advantage (Visser and Nemoto 2002) and the efficiency of the distribution network is regarded as a success factor for firms in the e-commerce market (Cho et al. 2008).

This topic has been tackled from many perspectives. Some authors have investigated the effects of B2C e-commerce logistics on the environmental impact of online purchasing processes (e.g. Edwards et al. 2011; Mangiaracina et al. 2015a; Mangiaracina et al. 2016; van Loon et al. 2015).

Some others have dealt with e-commerce logistics by adopting a general perspective, i.e. not focused on specific industries or markets, and sometimes also extended to the impacts e-commerce has on the whole supply chain (e.g. Zhou and Qi 2014). Others conversely have addressed this topic with reference to a specific context. For the purposes of our research, we have focussed on contributions related to the Chinese market. One of the most debated issues in this regard is the mismatch between the growth of the e-commerce market and the logistics one. Chinese e-commerce demand has grown extremely rapidly. However, logistics has become a bottleneck to further e-commerce development (Hou 2014) because it has not been able to keep up with this growth pace, slowed down by the time needed to develop the necessary infrastructure. In a sense, hence, e-commerce has brought a number of challenges to Chinese logistics that current players have not been able to face adequately (Hensher et al. 2015). Logistics services have often been affected negatively, with consequences such as late deliveries, order cancellation by merchants, lost goods in the shipping process, incorrect order fulfilment (Jiao 2014).

Despite the relation between e-commerce and logistics is acknowledged as critical, the issue seems to be treated mainly from a national perspective. More precisely, little relevance is given to the further complexities e-commerce can bring to logistics in a cross border environment. When coming to cross border e-commerce transactions the focus is more shifted towards linguistic barriers or marketing and branding aspects (e.g. Guercini and Runfola 2015; Gomez-Herrera et al. 2013; Lendle et al. 2012) while logistics seems disregarded.

A key interesting fact stemming from our analysis is therefore the lack of studies specifically dealing with the three main themes of our research (i.e. e-commerce, internationalisation and logistics) at the same time, although both international trade and e-commerce push for the construction of an effective and efficient distribution network. Correctly designing a distribution network is indeed a key element for the overall profitability of a firm (Mangiaracina et al. 2015b).

Given the gaps in the extant literature, the present paper intends to present an activity-based model to quantify the costs of different logistics solutions supporting an online sale abroad. More in detail, we will address this issue by answering the following research questions:

1. What are the main alternatives a company can choose from in order to configure the logistics of a cross border e-commerce sale?
2. How can these alternatives be modelled and what outcomes can be observed with reference to western fashion companies willing to sell to China?
3. What are the most important factors driving the choice of the optimal logistics solution?

3 Methods

The activities of the distribution process whose costs have been included in the model, which is the main output of this study, are (i) transportation, (ii) inventory management and (iii) handling:

- Transportation includes the entire delivery route from the country of origin to the Chinese consumer.
- Inventory management includes the inventory carrying costs of (i) the cycle stock, (ii) the safety stock, and (iii) the in transit stock.
- Handling refers to the operational costs to handle containers in the port/airport and in the warehouse/hub.

Another key decision factor is the service level (i.e. the time the Chinese consumer needs to wait for delivery after the online order is issued). The expected service level sometimes acts as a constraint. Indeed, if it is very short (e.g. two days) the fashion company will likely have no alternatives to using a warehouse in China in order to minimise the delivery time. For the purposes of our study, the service level has been considered as an embedded feature of each solution and explicated while describing the various logistics alternatives.

The model architecture consists of an inputs section with the information required to run the model, hidden spreadsheets for data computation (containing exogenous and endogenous data and the mathematical formulas) and an interface allowing the user to visualise the output graphs and tables referred to the cheapest solution, i.e. the one minimising total logistics costs.

The methodology used to build the model consists of three phases:

- Phase I: identification of the possible logistics alternatives supporting cross border e-commerce to China.
- Phase II: modelling of the logistics costs associated to the identified solutions.
- Phase III: application of the model to the fashion industry, performance of sensitivity analyses and validation of results.

The research approach is empirical, thus the model mainly relies on the information collected through interviews, case studies and a survey. Since our objective is to analyse logistics solutions already in use (or which could be used) to enable online export to China, our unit of analysis is the single case of a foreign manufacturing or retail fashion company operating online in China. Within the considered cases, we looked at the dyad "exporting company-logistics service provider", in order to better understand how logistics processes supporting cross border e-commerce are structured and how they can be modelled.

During Phase I, the authors conducted interviews with e-commerce service providers, fashion producers or retailers and logistics service providers operating globally and skilled in offering services supporting internationalisation to European firms. In total, 15 in-depth studies were derived from the interviews with the involved players. Additional information was then collected through a survey on a sample of 40 major Italian fashion producers and retailers that operate abroad through e-commerce. This process allowed deducing the possible logistics alternatives.

In Phase II data derived empirically from the survey, and the interviews were enriched with information taken from secondary sources (e.g. e-commerce websites, reports) and used as a basis to build the estimation model and the equations behind it.

In Phase III, the model was first tested by inserting input data from a heterogeneous set of fashion companies in order to calculate the logistics costs of the identified solutions. The information companies were asked to provide as endogenous inputs of the model are the following:

- Product information, e.g. weight and production costs.
- Logistics information, e.g. number of units per carton, number of cartons per pallet, location of the producer's warehouse, lead time to replenish the producer's warehouse.

The model also uses some embedded exogenous data as inputs, e.g. conversion rates, percentage of expected demand to allocate to Chinese cities, transportation fares.

Secondly, a sensitivity analysis was performed in order to show the impact of changing the annual demand and the value of the item on the optimal choice selection.

4 Results and Discussion

The authors have identified three solutions, pictured in Fig. 1, to deliver goods from a foreign producer or retailer's warehouse to the Chinese consumer's door (we hypothesise home delivery is adopted to fulfil an e-commerce order):

1. Solution "Courier" (C): distribution from a warehouse in the country of origin through express couriers.
2. Solution "Hub" (H): distribution through a sorting hub located in China.
3. Solution "Warehouse" (W): distribution from a warehouse in China.

It is important to note that the above-mentioned alternatives do not complete the set of admissible solutions. However, the selected options emerged as the most frequently used according to the interviewees and the respondents to the survey. Focussing on the most common logistics alternatives allows giving a reliable picture of reality while avoiding excessive complexity.

Some of the main characteristics and assumptions underlying each solution are listed below:

Solution C—This option implies the full outsourcing of the logistics activities. The courier takes the goods from the producer or retailer's warehouse in the country

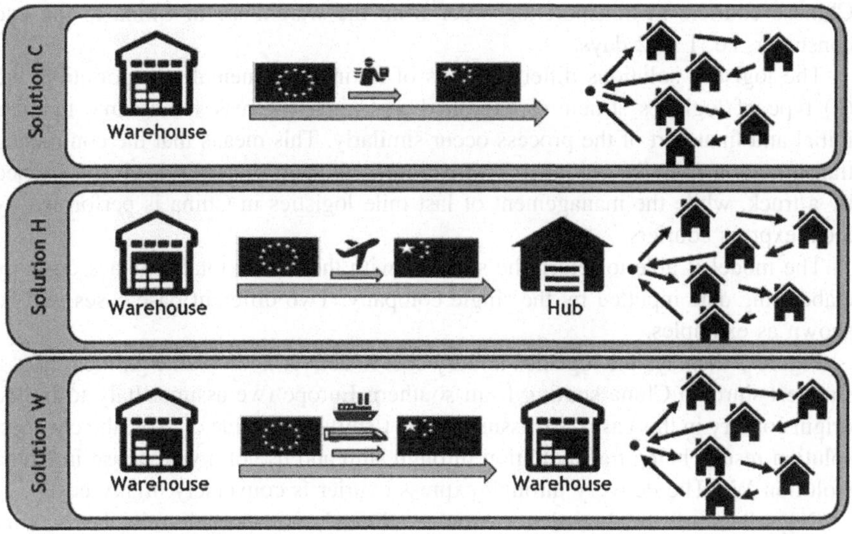

Fig. 1 Main logistics solutions supporting cross border e-commerce

of origin and fulfils the orders to the final customers. In this case, delivery lead times are on average of 3–5 days depending on the specific location of the customer. However, they could happen to be delayed a couple of days due to customs clearance procedures.

Solution H—We assume an intermediary consolidates orders from more producers until transportation of a full load is possible. The typical transportation mode for the intercontinental route is plane and the receiving hub is located in one of the Free Trade Zones (FTZ). These areas have recently been instituted to facilitate foreign firms to sell to Chinese customers. Thanks to their large logistics infrastructures, they are well equipped to supply logistics services. Shanghai's FTZ is the main hub for all the leading foreign carriers, such as Lufthansa Cargo, DHL and UPS and, therefore, it is served very frequently. In addition, administrative procedures are simplified and the goods benefit from the duty relief (Rafferty 2014; Quanlin 2015; Yao 2015; Zhang 2013). Delivering by plane takes around 5—8 days to conclude; in this case indeed, additional time needed to reach saturation of the transportation units before delivery starts might lengthen the process.

Solution W—We assume the intercontinental transportation takes place via ship or train and the Chinese warehouse is located in a FTZ. In this case, international transportation times are significantly higher (i.e. more than 40 days on average for the delivery by ship and around 20 days by train). However, having a warehouse easily allows managing in advance the shipping process so that goods are already available in China when needed. Therefore, in this case delivery lead times to the final customer should be the lowest of any alternatives, as goods are physically in China when the online purchase order is submitted. This allows improving the service level. In this case the delivery time corresponds to the time needed by

Chinese couriers to transport the order from the warehouse in China to the end consumer, i.e. 1 or 2 days.

The logistics solutions differ in terms of (i) intercontinental transportation and (ii) type of logistics structures exploited in China, whereas we assume that the initial and final part of the process occur similarly. This means that the continental transportation from the producers' warehouse to the port or airport is always carried by a truck, while the management of last mile logistics in China is performed by local express couriers.

The model is able to return the solution with the lowest total logistics costs by elaborating data inputted by the single company. Two different base cases will be shown as examples.

Figure 2 displays the logistics unitary cost to deliver a low-value garment (e.g. a simple t-shirt) to China starting from southern Europe (we assume Italy to be the origin country in this case). By assuming the t-shirt has a value of 10 €, the cheapest solution results being transportation through ship and use of a warehouse in China (solution W). The delivery through express courier is conversely highly costly.

Regardless of the logistics solutions adopted, transportation is the activity weighting the most, especially in solution C where it accounts for 99% of the total logistics cost. The only exception happens in solution W with the use of the ship, where transportation accounts for 33% of logistics costs and inventory carrying costs weight 42% over the total.

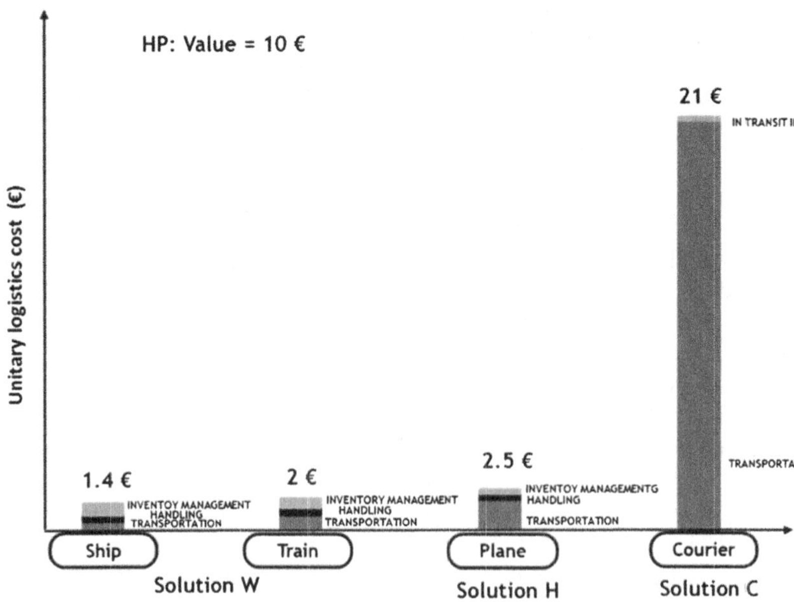

Fig. 2 Logistics costs of a low-value garment

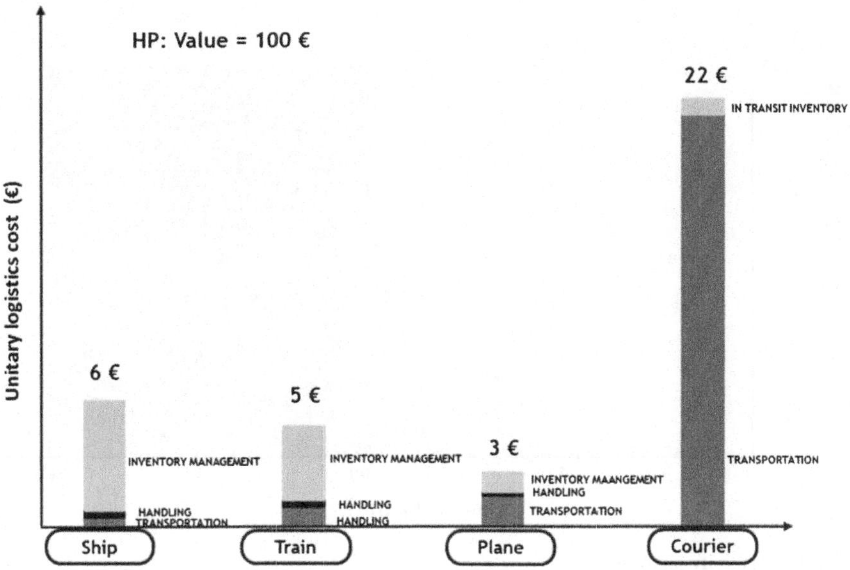

Fig. 3 Logistics cost of a high-value garment

By considering a fashion item of higher value, e.g. a garment worth 100 €, the cheapest solution is conversely the second one, involving the presence of a hub in China and the use of the plane for transportation (solution H) as pictured in Fig. 3. In this case it is possible to note that inventory management costs have a major impact on solution W due to the higher value of the item, which make this alternative less efficient than the one considering the hub. Solution C is again the most expensive. Moreover, despite the increasing value of the product, the cost of the express courier varies negligibly with respect to the previous case simply because courier charges generally depend on the weight of the item rather than its value. Courier fares are also negatively related to the quantity to be sent, i.e. the larger the volumes, the lower the fare.

This evidence suggests the value of the product as well as the annual traded volumes have an impact on the selection of the most efficient logistics solutions. This is the reason why a sensitivity analysis was performed to measure the effects of variations in both these parameters. Figure 4 depicts the results of the sensitivity analysis.

In coherency with the characteristics of the two services, solution W dominates for low value items (less than 45 €). The lower the value of the item the more convenient is to internationally transport by ship rather than by train.

As the product's value grows the convenience to ship by plane and use a hub in China (solution H) increases. More specifically, this solution dominates when the value of the item is higher than 90 € and for lower demand levels. This happens for a number of reasons, briefly summarised below:

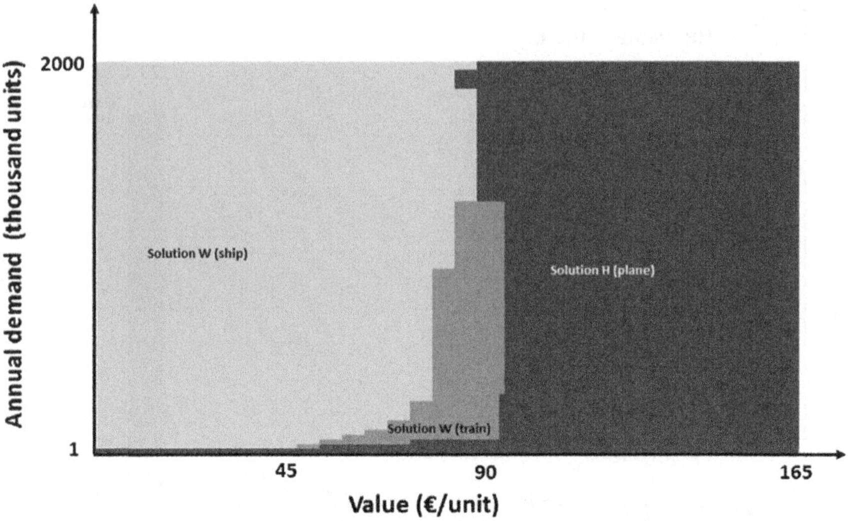

Fig. 4 Sensitivity analysis

- The lead-time to transport goods to China is lower than the other solutions' one. For this reason, the impact of in transit stock decreases when airfreight is employed. The impact of this decrease grows with the item's value.
- The inventory carrying costs decrease because no warehouse is present in China. Solution H only assumes there is a warehouse in the country of origin, thus there is no duplication of stock as we have a distribution network with only one echelon. On the contrary, solution W implies the presence of two echelons and inventory is duplicated. Furthermore, the replenishment time to refill the Chinese warehouse (solution W) is both longer (due to the transportation lead time) and more uncertain (because of custom clearance issues).
- This solution appears to be optimal also when demand volumes are very low because of the FTZ warehouse costs. Indeed both the rent and the picking costs have a minimum daily threshold that has to be paid anyways. When the annual demand is very low and the sum of the unitary daily costs per unit of space rent and picking do not reach the minimum daily cost, the impact of the rent and picking cost on the single unit is very high and the solution including the Chinese warehouse falls out of the convenience zone.

As far as the demand level is concerned, we can see it is determinant especially in case the intercontinental transportation is conducted by train (within solution W). The train has indeed a modest capacity (with respect to the ship) and is therefore suitable for more limited volumes.

It is furthermore interesting to note how solution C is always dominated in this model. This happens because the algorithm behind the model, in its current version, can only take into account costs related to the logistics activities listed in the

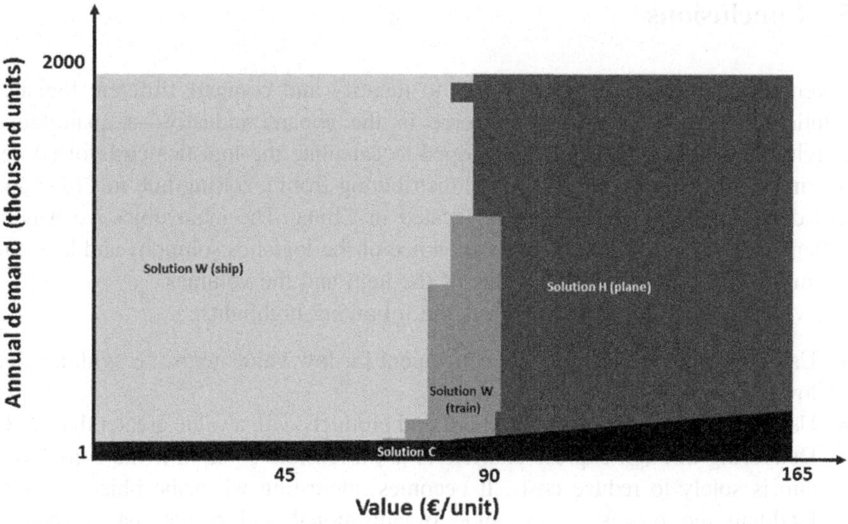

Fig. 5 Hypothetic outcome of a model considering reductions in the operational complexity of managing cross border logistics

methods section. The model does not consider the "operational effort" needed by a company to manage the logistics process to such a distant country directly.

This variable is hard to quantify, however it surely plays a major role in the choice of the distributive solution. Moreover, solution C would probably be the one reducing such operational complexity the most because the logistics process is completely outsourced.

The authors guess that, should this variable be included, the optimal solution matrix would look similar to that in Fig. 5. That is to say, there must be an area of convenience for solution C if the company's objective is not to simply minimise the total logistics cost, but to reduce the complexities of directly managing this process. We suggest this solution could be optimal for limited volumes but increasingly with the value of the garment.

The sensitivity analysis reveals the most important factor driving the choice of the logistics solution is the value of the item: the higher the value the greater the convenience to opt for the solution with a hub in China due to the high incidence of storage costs occurring in case a Chinese warehouse is owned. The expected demand also affects the choice but to a lower extent since it seems to have an impact only for medium value items. In addition, the trade-off between operational simplification and cost reduction is clearly an influencing factor. Results demonstrate that solution C is never the most efficient. Its selection is therefore justified by reasons other than cost optimisation, i.e. the reduction of operational complexity reached through complete externalisation of the logistics activities. Eventually, as already highlighted while presenting the model, the service level plays a major role in the decision process as it acts as a constraint of the single solution.

5 Conclusions

Given the purpose of this paper—i.e. to identify and compare different logistics solutions for cross border e-commerce in the apparel industry—a quantitative activity-based model has been developed to calculate the logistics costs of (i) delivering through express couriers, (ii) distributing from a sorting hub in China and (iii) distributing from a warehouse located in China. The main objective was to understand what factors influence the choice of the logistics solutions and how the optimal choice varies with the value of the item and the volumes.

Evidence from this paper suggests the following highlights:

- Using a warehouse in China is convenient for low value items, i.e. with a value lower than 45 €.
- Using a sorting hub is worth for high-end products with a value greater than 90 €.
- Delivering through express couriers is not a feasible solution if the company's aim is solely to reduce costs. It becomes interesting when the objective is to facilitate the process since some organisational and operational burden is outsourced.
- Transportation plays a major role in driving logistics costs upwards due to the long distances. This especially holds true for low value garments. If the value of the item increases then also the warehousing costs become relevant, that is also why the solution with a hub (solution H) in place of a warehouse (solution W) becomes more convenient. Handling seems not to have a relevant impact on logistics costs in any of the examined cases.
- In terms of timings, having a warehouse in China significantly reduces the delivery time to the customer. Carefully planning the replenishment of the warehouse is a key for such solution to guarantee high service level.

The present paper has significant academic and practical implications. From a theoretical point of view it aims to originally contribute to the extant literature by addressing three topics that have typically been treated separately, i.e. logistics, e-commerce and internationalisation. Another strength point is the quantitative nature of the model which calculates costs that can be compared. This provides merchants and retailers in the fashion industry with an effective tool to learn more about the logistics alternatives to serve the promising Chinese B2C e-commerce market and identify the most efficient strategy for them.

Last, the model architecture is very flexible: it could be extended easily to other destination countries and industries if new specific data were collected.

However, this model presents some limitations. First, the results currently obtained are not sufficient to derive a comprehensive evaluation of the logistics alternatives supporting cross border e-commerce to China as they mainly consider cost minimisation as a selection driver. In reality, other reasons may drive business decisions. Another limitation is the fact that returns management is not included in the analysis, despite it is relevant for e-commerce based sales in the fashion industry.

Based on the identified limitations, the following streams for future research are suggested:

- Including the return management among the considered logistics activities.
- Elaborating a more complex version of the model where also the benefits of a solution (not only the costs) are computed so that the evaluation of the alternatives can be more complete.

References

Bonetti F (2014) Italian luxury fashion brands in China: a retail perspective. Int Rev Retail Distrib Consum Res 24(4):453–477

Cho JJ, Ozment J, Sink H (2008) Logistics capability, logistics outsourcing and firm performance in an e-commerce market. Int J Phys Distrib Logist Manage 38(5):336–359

Delfmann W, Albers S, Gehring M (2002) The impact of electronic commerce on logistics service providers. Int J Phys Distrib Logist Manage 32(3):203–222

E-commerce Foundation (2015). Global B2C E-commerce report 2015

Edwards JB, McKinnon AC, Cullinane SL (2011) Comparative carbon auditing of conventional and online retail supply chains: a review of methodological issues. Supply Chain Manage Int J 16(1):57–63

EMarketer (2015). China Ecommerce: 2015 Market Update

Forrester (2015). Forrester Research Online Retail Forecast, 2014–2019 (Asia Pacific)

Ghezzi A, Mangiaracina R, Perego A (2012) Shaping the e-commerce logistics strategy: a decision framework. Int J Eng Bus Manage 4(3):1–13

Ghezzi A, Georgiades M, Reichl P, Le-Sauze N, Cairano-Gilfedder CD, Mangiaracina R (2013) Generating innovative interconnection business models for the future internet. Info 15(4): 43–68

Gomez-Herrera E, Martens B, Turlea G (2013) The drivers and impediments for cross-border e-commerce in the EU. European Commission Joint Research Centre, Institute for Prospective Technological Studies

Guercini S, Runfola A (2015) Internationalization through e-commerce. The case of multibrand luxury retailers in the fashion industry. In: Stöttinger B, Schlegelmilch BB, Zou S (eds) International marketing in the fast changing world, advances in international marketing, vol 26. Emerald Group Publishing Limited, pp 15–31

Hensher DA, Zhang Z, Rose J (2015) Transport and logistics challenges for China: Drivers of growth, and bottlenecks constraining development. Road Transp Res 24(2):32–41

Hou XF (2014) Analysis on the situation of China's E-commerce logistics. Adv Mater Res 926–930:3750–3753

Jiao Z (2014) Development of e-commerce logistics in China. In: Liu B et al (ed) Contemporary logistics in China. Consolidation and deepening. Springer

Lendle A, Olarreaga M, Schropp S, Vézina PL (2012) There goes gravity: how ebay reduces trade costs. CEPR discussion paper no DP9094

Mangiaracina R, Marchet A, Perotti S, Tumino A (2015a) A review of the environmental implications of B2c ecommerce: a logistics perspective. Int J Phys Distrib Logist Manage 45 (6):565–591

Mangiaracina R, Song G, Perego A (2015b) Distribution network design: a literature review and a research agenda. Int J Phys Distrib Logist Manage 45(5):506–531

Mangiaracina R, Perego A, Perotti S, Tumino A (2016) Assessing the environmental impact of logistics in online and offline B2c purchasing processes in the apparel industry. Int J Logist Syst Manage 23(1):98–124

Premazzi K, Castaldo S, Grosso M, Hofacker C (2010) Supporting retailers to exploit online settings for internationalization: The different role of trust and compensation. J Retail Consum Serv 17(3):229–240

Quanlin Q (2015) Guangzhou charts ambitious plan to become shipping hub. China daily. http://usa.chinadaily.com.cn/epaper/2015-02/07/content_19530889.htm. Accessed 20 Dec 2015

Rafferty T (2014) The Shanghai free trade zone. The economist

Ramanathan R, George J, Ramanathan U (2014) The role of logistics in e-commerce transactions: an exploratory study of customer feedback and risk. In: Ramanathan U, Ramanathan R (eds) Supply chain strategies, issues and models. Springer, London, pp 221–233

van Loon P, Deketele L, Dewaele J, McKinnon A, Rutherford C (2015) A comparative analysis of carbon emissions from online retailing of fast moving consumer goods. J Clean Prod 106: 478–486

Visser JGSN, Nemoto T (2002) E-commerce and the consequences for freight transport. In: Taniguchi E, Thompson RG (eds) Innovations in freight transport. WIT Press, Boston, pp 165–193

World Trade Organisation (2015) International trade statistics 2015. http://www.wto.org/statistics. Accessed 7 March 2016

WorldPay (2015) Global payment report. https://www.worldpaymentsreport.com. Accessed 9 Feb 2016

Yao R (2015) The new free trade zones explained, Part I: Guangdong, China Briefing. http://www.china-briefing.com/news/2015/04/23/new-free-trade-zones-explained-part-guangdong.html. Accessed 20 Dec 2015

Zhang S (2013) Import-export taxes and duties in China, China Briefing. http://www.china-briefing.com/news/2013/03/11/import-export-taxes-and-duties-in-china.html. Accessed 20 Dec 2015

Zhou Y, Qi J (2014) Supply chain design strategy based on e-commerce. Appl Mech Mater 457–458:1411–1414

Author Biographies

Maria Giuffrida is a Researcher of the Export Observatory at Politecnico di Milano School of Management and a PhD student in the Department of Management, Economics and Industrial Engineering of Politecnico di Milano. In 2014, she received her Master's Degree in International Management cum laude from Bocconi University, Milano, Italy. Her main research interests include global operations and supply chain management, export models, risk analysis and risk management in global environments.

Riccardo Mangiaracina is an Assistant Professor at Politecnico di Milano, Italy, where he lectures and undertakes research on logistics, supply chain management and e-commerce. He is the Director of the B2C E-Commerce and Export Observatories of Politecnico di Milano School of Management. His current research interests include B2C e-commerce, logistics, supply chain management and sustainability, and intelligent transportation systems. He published a number of contributions in the field of e-commerce and new technologies, including papers in the International Journal of Physical Distribution and Logistics Management, and the Journal of Internet Banking and Commerce.

Alessandro Perego is a Full Professor of Logistics and Supply Chain Management at Politecnico di Milano, where he chairs the course of Logistics and Production Systems Management and Logistics management. During his career, he chaired also the courses E-operations and Industrial Logistics. He is the co-Founder and the Scientific Director of the Observatories Digital Innovation of the Politecnico di Milano School of Management. He is Director of the Department of Management, Economics and Industrial Engineering.

Angela Tumino is an Assistant Professor and currently lectures on Logistics and Production Systems Management at Politecnico di Milano, Italy. She is the Director of the Internet of Things Observatory of Politecnico di Milano School of Management. Her main research interests include logistics, supply chain management and sustainability, value assessment, information and communication technologies (ICT). She is the author of several journal papers and conference proceedings, including contributions in the International Journal of Operations & Production Management, Supply Chain Management: An International Journal, International Journal of Production Economics, and International Journal of Advanced Manufacturing Technology.

Development of Scheduling Systems for a Shoe Factory Through IDEF0 and RFID Technologies

Maurizio Bevilacqua, Filippo E. Ciarapica and Giovanni Mazzuto

Abstract In recent years, an increased emphasis on the use of operations management (OM) models and theories has emerged, due to the strict links between firm competitive and supply chain strategies. Companies have thus to choose the appropriate and specific responses so that to face the market challenges. In this paper, an optimisation process for the footwear production is discussed through the IDEF0 approach, reengineering the production planning task and introducing Radio Frequency Identification (RFID) tools to better map the production process. The AS-IS map of a famous footwear brand has been analysed in order to identify its strengths and weaknesses through a WHAT-IF analysis. Then, the TO-BE map has been identified introducing a new production scheduler as well as RFID technologies. The analysis highlighted that the new management approach can help the company yielding a better production management, increasing thus the opportunity to expand on the global market. In particular, RFID provides a solution to difficult logistical tracking of inventory or equipment, particularly in applications where optically based systems fail and when read/write capabilities are required. Moreover, investment economic feasibility analysis is described showing the benefit of the reengineered management system.

Keywords Business process reengineering · What-if analysis · IDEF0 · RFID · Optimisation

M. Bevilacqua · F.E. Ciarapica (✉) · G. Mazzuto
Industrial Engineering and Mathematical Science Department, Marche Polytechnic University, Via Brecce Bianche, 60131 Ancona, Italy
e-mail: f.ciarapica@univpm.it

M. Bevilacqua
e-mail: m.bevilacqua@univpm.it

G. Mazzuto
e-mail: g.mazzuto@univpm.it

R. Rinaldi and R. Bandinelli (eds.), *Business Models and ICT Technologies for the Fashion Supply Chain*, Lecture Notes in Electrical Engineering 413,
DOI 10.1007/978-3-319-48511-9_15

1 Introduction

In recent years, the use of operations management (OM) models and theories has been stressed due to the strict links between firm competitive and supply chain strategies. Companies have thus to choose the appropriate and specific responses so that to face the market challenges. One of the most relevant manufacturing sector experiencing such changes is the footwear sector, particularly relevant in economic Italian context. In more detail, and referring to 2014, the Italian footwear sector counted 5.031 companies with 76.610 employees, representing one of the most important sector in the fashion system (data available on line at http://www.assocalzaturifici.it/ 01/02/2016).

Being proactive i.e. anticipating possible market demands changes, requires an organisation to better manage the production process highlighting bottlenecks and evaluating possible implementations of new technologies. In such a complex contex many companies are thus increasingly investing in wards of "process analysis and re-engineering" increasing their capabilities in terms of prompt supply chain strategy implementation, Total Quality Control and Concurrent Engineering.

Computerised stitching introduction improves process efficiency, automated leather "roughing" i.e., the removal of leather top surface, lowered highly skilled labour requirement. Injection moulding, enabling mass production of items such as synthetic soles and heels and computerized water jets cut by replacing manual processes represent other examples of significant developments.

It is important to highlight that manufacturing facilities are complex, dynamic and stochastic systems. Operation Management tools and techniques strongly help managers in the field of production control as well as in labour resource management. As far as production schedules are taken into account, it is worth to remind that a production schedules coordinate activities in order to increase productivity and minimise the operating costs. A production schedule could identify resource conflicts, define and control the release of jobs to the shop, ensuring so that required raw materials are supplied in time, and identifying time intervals for proper maintenance activities. Moreover, information technology can be usefully applied to match market requirements. According to LaForge and Craighead (1998) it is possible to conclude that computer based scheduling can help manufacturers to respect on-time delivery, to quickly respond to customer orders, and to create realistic schedules, with a key success factor being the use of finite scheduling techniques and their integration with other manufacturing planning systems and technologies.

The paper is organised as in the following: Sect. 2 reports a brief description of the research approach related to IDEF0 and RFID technology; Sect. 3 describes the analysed case and Sect. 4 discusses the obtained results highlighting the economic point of view. Finally, Conclusion summarizes and describes the paper outcomes and the benefits of the proposed approach.

2 The Research Approach

In order to optimize the production process through the adoption of an appropriate scheduler, the main problems of the production system have been investigated. IDEF0 diagrams and maps have been used so that to identify the of information flow characterizing the AS-IS situation of the case study.

The analysis described in the following aims at defining an efficient strategy for production control. To this extent production data collection and analysis allows to clearly define the production steps schedules, keeping also into account for the constrained resource availability.

The analysis of RFID technological solution and their suitable adoption allows a real time production progress data to be gathered, ensuring thus a better production process planning and re planning. The TO-BE IDEF0 diagram described in the following shows the future process map, highlighting the benefits arising due to the introduction of the scheduler and of RFID devices.

2.1 *The IDEF0*

The Integration Definition Function Modelling (IDEF0) is a modelling language, with graphical and text capabilities, designed to model the decisions, functions and activities of systems and organizations. IDEF0 derived from the graphical language SADT (Structured Analysis and Design Technique) developed in 1972 (Marca and McGowan 1988). IDEF0 is now an IEEE Standard (KBSI 1998). The IDEF0 has a hierarchical structure. The hierarchical structure allows an overall view of the system also giving the representation of any portion in detail. A top-level context diagram (the A00 diagram) is composed by a single activity box and identifies the model context, the model purpose and the viewpoint. The A0 context diagram is then decomposed into its main sub-activities through a child diagram. Each sub-function can be again decomposed into low-level child diagrams. Such a hierarchic decomposition is schematised in Fig. 1.

Moreover, IDEF0 helped to implement Process Reengineering in different sectors in addition to the industrial one: agricultural (Papajorgji and Pardalos 2009), bank service (Climent et al. 2009), emergency management (Bevilacqua et al. 2012) etc.

2.2 *The RFID Technology*

Radio frequency identification (RFID) is an emerging technology that has great potential for logistics, supply chain management and quick response systems (Uckelmann et al. 2012). Chao et al. (2007) identified RFID as one of the ten

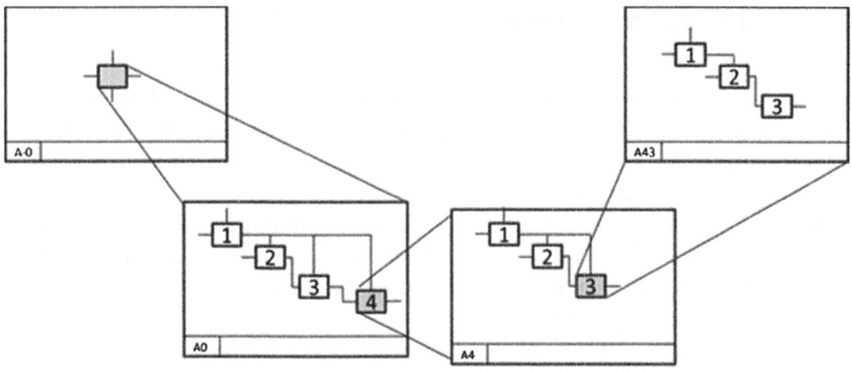

Fig. 1 IDEF0 hierarchic structure

greatest contributory technologies of the 21st century. This technology has been adopted by a variety of enterprises to achieve cost-savings, to increase efficiency and to gain a competitive advantage (Bilge and Ozkarahan 2004). From a literature analysis, supply chain management, SCM (Michael and McCathie 2005), health industry, and privacy issues emerge as the major trends in RFID application. RFID provides tracking information without requiring direct contact with the object being tracked. This is an important feature for applications like shoplifting deterrence, vehicle identification, and animal or people monitoring. RFID refers to technologies and systems that use radio waves (wireless) to transmit and uniquely identify objects (Finkenzeller 2003; Van der Heijden 2006). Many authors have discussed the expected impact of RFID technology and the EPC Network on supply chain dynamics and their related cost savings. It is important to highlight that quantitative assessments of the economical results of their adoption are limited (Bottani et al. 2010).

3 The Case Study

The analyse company has a steady production capacity of 2,000 shoes pairs per day, being now utilized up to 50%. The current workforce amounts to 161 peoples including 42 clerks, 115 workmen and 4 collaborators. The company controls the entire value chain: from market research to the product design, and from materials research to the production; ending with the sale and distribution phases. The firm production is 100% Made in Italy.

The manufactured products, classified as fine and luxury type, are addressed to a target attentive to detail, finishes, materials research, preciousness and exclusivity. The firm end product is a very high quality one, with a retail price of starting from 250€ up to 1,000€ per pair.

The production, as well as the products marketing, is completely internally managed. Firm production is directed to a niche market, focusing on high quality products with a strong personalization. The firm reference market is characterised by selected customers with a relevant purchasing power. The firm main markets slightly undergo economic crisis, being instead very sensible to fashion and luxury emerging trends as well as to product innovations. Products design is thus resulting from a careful marketing analysis focussing on consumer trends and desires, a customer added value being the fully in-house style development.

4 Results and Discussion

The production cycle AS-IS map is shown in Fig. 2. When a customer order is received, it is transferred to the enterprise information system and, as soon as the customer payment is received, the production process starts with the proper leather selection. Leather is then transferred to the cutting department and once cut, the shoes production can finally start, with the sewing, pulling over, chain stitching, outsewing, edge grinding, coating applying and finishing step fulfilled in sequence. A quality control on the semi-finished product takes place at the end of each phase. For each production phase, relevant data are collected so that to feature the phase itself. The production phase dataset has then been analysed to obtain lead time descriptive statistics for both single phase and whole production process. To more strictly control work in progress evolution and assembly line takt time a simple scheduler has been introduced based on input output control logic.

The introduction of such a scheduler brings considerable benefits from an organizational point of view as:

- Products due date respects, suggesting corrective measure in case of unexpected delays;

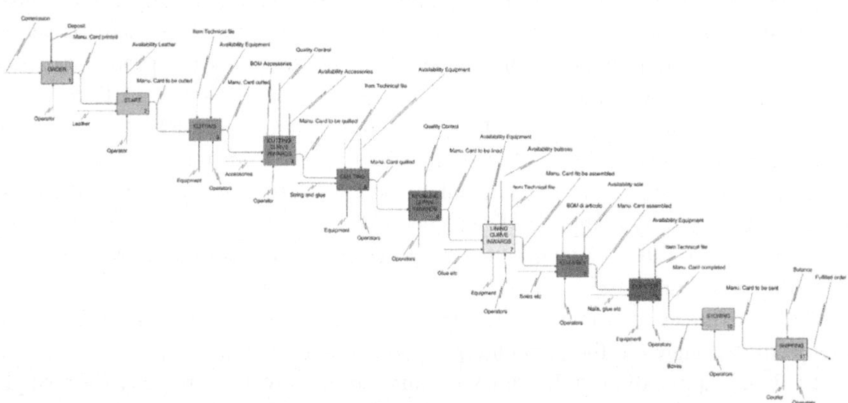

Fig. 2 The AS-IS scenario of the production cycle

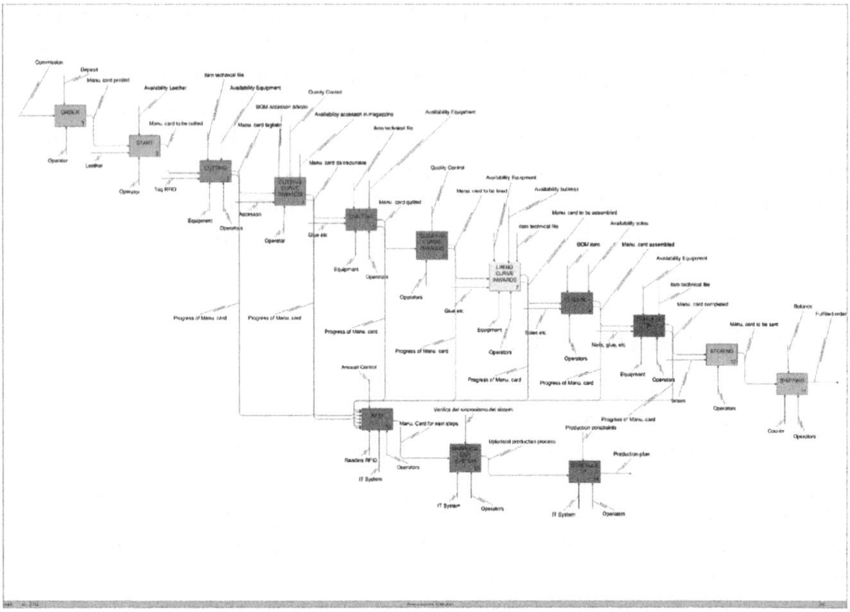

Fig. 3 The TO-BE scenario of the production cycle

- A more tight and rigorous production phase planning along the process;
- A suitable flexibility level to account of customer order changes;
- A realistic resource level utilisation and control along the various production phases.

It is also important to point out that, the scheduler can ensure greater benefits if coupled with and RFID system, so that to allow a process real-time data acquisition. RFID technology implementation also allows a productivity increase especially due to the cut of operator time spent in subcontract work control, actually valued equal to one hour per day. Such a time saving, with the operator time used for added value tasks, can thus improve process productivity lowering the production lead time. Figure 3 shows the process future map, i.e. the IDEF0 TO-BE scenario, taking into account the adoption of the scheduler coupled with RIFD systems.

An economic feasibility study has also been carried out for the RFID system acquisition and installation. The scheduler cost has not been reported since it will be internally realised by company IT staff people.

Tables 1 and 2 show a detail of the proposed RFID cost, for both investment and operating cost items.

The investment costs include the tags and readers purchase, their installation and the needful changes in the firm enterprise information systems.

RFID tag application in the shoes is valued to require 1.5 h per day for a single operator, whose labour cost is equal to 20 €/h. Table 3 summarises the proposed investment annual income due to the cut of subcontract work control (items manual

Table 1 Investment costs

Cost item	Cost	Total cost (€)
Reader RFID	1.500 € × 8 readers	12.000
Installation	8.000 €	8.000
Programming	20.000 €	20.000
RFID tag	0.10 €/tag	50.000
Total		90.000

Table 2 Operating costs

Management cost	Cost	Amount	Annual cost (€)
Maintenance tag	5.000 €	annual	5.000
Maintenance reader	2.000 €	annual	2.000
Tag application	20 €/h	1.5 h/day	8.550
Total			15.550

Table 3 Income from the investment

Income	Margin (€/h)	Amount (h/year)	Annual income (€)
Item manual counting	20	570	11.400
Production progress	20	1140	22.800
Inventory	20	256	5.120
Total			39.320

counting, now 570 h/year), automated production progress control (4 operators for 16 h not less than 4 times a year, now 1140 h/year) and inventory control (now 256 h/year).

5 Conclusion

The paper has analysed the introduction of IT based technologies for a better control on a luxury shoes production system. The proposed solution can be summarised with the introduction of RFID systems coupled with a novel scheduler based on input output control principles. The real time shop floor data acquisition helps to fully respect the orders due date, checking the production progress and also ensuring suitable process flexibility in terms of resource utilisation, due to customer order changes management. RFID system introduction can help the company to implement more effective production planning procedures as well as to raise productivity, precondition for a possible market growth.

References

Bevilacqua M, Ciarapica FE, Paciarotti C (2012) Business process reengineering of emergency management procedures: a case study. Saf Sci 50(5):1368–1376

Bilge B, Ozkarahan I (2004) Strategic tactical and operational production-distribution models: a review. Int J Technol Manage 28:151171

Bottani E, Montanari R, Volpi A (2010) The impact of RFID and EPC network on the bullwhip effect in the Italian FMCG supply chain. Int J Prod Econ 124(2):426–432

Chao CC, Yang JM, Jen WY (2007) Determining technology trends and forecasts of RFID by a historical review and bibliometric analysis from 1991 to 2005. Technovation 27:268–279

Climent C, Mula J, Hernández JE (2009) Improving the business processes of a bank. Business Process Management Journal 15(2):201–224

Finkenzeller K (2003) RFID handbook: fundamentals and applications in contactless smart card identification, 2nd edn. Wiley

KBSI (1998) IEEE standard for functional modeling language-syntax and semantics for IDEF0. IEEE Std 1320.1–1998. IEEE

LaForge RL, Craighead CW (1998) Manufacturing scheduling and supply chain integration: a survey of current practice. American Production and Inventory Control Society, Falls Church, Virginia

Marca DA, McGowan CL (1988) SADT: structured analysis and design technique. McGraw-Hill Book Co, Inc: New York, NY

Michael K, McCathie L (2005) The pros and cons of RFID in supply chain management International conference on mobile business, 623–629

Papajorgji PJ, Pardalos PM (2009) Advances in modeling agricultural systems. Springer optimization and its applications, Vol 25, No. 10, pp 172

Uckelmann D, Scholz-Reiter B, Rügge I, Hong B, Rizzi A (2012) Communications in computer and information science: preface. Commun Comput Inf Sci 282 CCIS, 5–6

Van der Heijden H (2006) Mobile decision support for in-store purchase decisions. Decis Support Syst 42:656–663

Proposal of a Multi-method Decision Support System for the Fashion Retail Industry

Giada Martino, Marcello Fera, Raffaele Iannone and Salvatore Miranda

Abstract Fashion and Apparel (F&A) market, characterized by fast changes in trends and demand, by short product life-cycles and by broad assortments, requires a responsive/demand driven Supply Chain (SC) focused on products availability, real-time information sharing and speed in matching customers requests (Iannone et al. 2013). In this context, the presented paper shows the results of a three-year research project by firstly analysing the overall structure and characteristics of a traditional SC in this sector Iannone et al. (2015) in order to identify the most critical aspects and processes. From a risk analysis it emerged that the correct Time Management, intended as the ability of being responsive to market fluctuation, is the most critical target for fashion business. In this context, the presented work proposes a reference framework for the definition and subsequent optimisation of the physical and informative flows, which is based on a deviation analysis of demand and an adjusting feedback loop. In last years, the wide spread of e-commerce and mobile purchasing is deeply changing the retailing industry leading companied to adopt a new integrated strategy, called Omni-Channel Retailing. The management of both physical and mobile channels not simply means managing an additional on-line demand but requires the actual integration of all the processes of planning and execution in order to optimize performances. With these perspectives, the proposed framework has been revised and extended in order to represent a company implementing this new strategy and, after the definition of a suitable set of Key Performance Indica-

G. Martino · R. Iannone (✉) · S. Miranda
Department of Industrial Engineering, University of Salerno,
via Giovanni Paolo II 132, 84084 Fisciano (SA), Italy
e-mail: riannone@unisa.it

G. Martino
e-mail: gmartino@unisa.it

S. Miranda
e-mail: smiranda@unisa.it

M. Fera
Department of Industrial and Information Engineering,
Second University of Naples, Via Roma 29, 81031 Aversa (CE), Italy
e-mail: marcello.fera@unina2.it

R. Rinaldi and R. Bandinelli (eds.), *Business Models and ICT Technologies for the Fashion Supply Chain*, Lecture Notes in Electrical Engineering 413,
DOI 10.1007/978-3-319-48511-9_16

tors (KPIs), allowed us to evaluate how it may impact on the performances of a traditional SC. The integration of all these analysis and the correct evaluation of the defined set of KPIs may represent a useful system for supporting fashion companies in the strategic decision making process.

Keywords Supply chain management · Fashion and apparel industry · Omni-channel retailing · Optimisation · Key Performance Indicator · Simulation

1 Introduction

The era when it was acceptable to manage twelve to eighteen months product lifecycles in the fashion industry is over (Rosenblum 2015). The F&A Industry is, in fact, one of the key pillars of the global economy, but it also represents one of the most flexible and unpredictable Industries, given the high volatility of demand and fast changes in customer tastes and trends. Until the late 1980s, traditional F&A retailers used demand forecasting for the definition of their operations plans long before the actual time of product consumption, i.e. the sales season. In the last decades, though, this sector experienced a real revolution due to the introduction of the "Fast Fashion" model. This retail strategy is based on adapting merchandise assortments to current and emerging trends as quickly and effectively as possible. Fast fashion retailers have, in fact, replaced the traditional push model—based on the prediction of upcoming trends by designers—with a pull approach, in which retailers respond to customer demand trying to get the right product in the market within just a few weeks, versus an industry average of six months (Hansson 2011). Besides this deep change in the management model, the whole Retail business—not only the Fashion and Apparel one—is now facing a new tremendous challenge: surviving in a model that includes the Internet (Shoenbachler and Gordon 2002). As online purchasing is continuing to grow, the future of pure brick-and-mortar retailers is called into question. Despite physical stores still remain at the heart of the customer relationship, in fact, online and mobile sales appear to rule (Brown et al. 2013). Then, in the last decades, the major retailing companies are converging towards the "Omni-Channel" strategy, i.e. a synchronized operating model in which all of company's channels—traditional stores and mobile channels—are aligned allowing companies to meet customers' requirements and to be more competitive. In other words, connected customers can shop for and purchase the same items across many different channels: in a physical store, on their home or laptop computers, on their connected mobile devices (Motorola Solutions 2012). In this complex and dynamic context, the presented work illustrates the results of a three-year research project and has the main aim of proposing a tool to support companies of this sector in the decision making process of definition of the operations plans. To achieve this result, first part of the research project is focused on the definition of the main risks connected to the whole SC flow (Sect. 2). This Risk Analysis was considered strategic for the design of the proposed approach and allowed us to identify the main target that managers try to perceive and the most critical areas and processes. In particular, it emerged that main aim of fashion business is the correct *Time Management*, intended as the ability of being responsive to market fluctuation. This problem is analysed in Sects. 3 and 4 by proposing a framework for the optimisation of performances for the traditional pure brick-and-mortar case, and then for the Omni-Channel strategy. In the end, Sect. 5 analyses and discusses main results obtained from the research project.

2 Supply Chain Risk Assessment: Understanding Main Targets and Critical Areas

As already mentioned, the research project starts with a detailed risk analysis for the identification of the main issues and criticalities and for the appropriate definition of the Operations strategy. According to the United States, in fact, Fashion Industry Association (Sheng 2014), the second top ranked fashion industry challenge is "managing supply chain risks" immediately after "increasing production or sourcing costs". The increasing trend towards globalization and outsourcing is, in fact, leading many industrial sectors to entrust relevant parts of their business to suppliers often located in developing countries. This phenomenon is causing loss of control and of full visibility of the supply chains, thus increasing risks connected to any possible change or disruption. In the highly dynamic context of the F&A Industry, the proper assessment and management of the Supply Chain risks can be crucial for its efficiency. The main risk factors of a traditional Fashion Supply Chain were identified by Martino et al. (2015), according to the following targets: Cost Reduction, Market Driven Orientation, Brand Internationalization & Market Expansion and Environmental Sustainability. These factors were grouped into the following homogeneous clusters:

1. *Competitive environment*: includes external risks not directly controllable by main company, related to the target market and to competitors;
2. *Relationship between SC actors*: concerns policies and strategies with which all actors interacts;
3. *Offer*: this cluster is specifically related to offered products and services;
4. *Informative flow*: involves all factors connected to communication, informative systems and data exchange;
5. *Process and timing control*: it is focused on management and optimisation of processes;
6. *Distribution*: involves transports and material handling activities;
7. *Knowledge of customers/end-users*: refers to the ability to appropriately capture trends and market needs;
8. *Quality and quantity control*: it is focused on the production processes.

Then, thank to a deep analysis of both current practice (involving 12 managers from 6 different companies operating in the Italian Fashion and Apparel Industry) and literature (involving more than 60 papers from the most important journals deal-

Table 1 Prioritization of the targets according to their weights

Priority	Target	Sub-target	Weight
1	Cost Reduction	Time to Market	0,4447
		Mat. and Inf. flow management	0,3137
2	Market Driven Orientation	Improvement in market sensitivity	0,1672
		Brand Attractiveness	0,0628
3	Brand intern. & market expansion		0,0087
4	Environmental sustainability		0,0029

ing with the topics of supply chain risk management, retailing and operations management in the F&A Industry) and through the use of the Analytic Network Process (ANP) approach (Saaty and Vargas 2006), a priority list is defined. This list highlighted the most crucial area on which to focus the attention. Results obtained are reported as follows:

- Table 1 shows the targets ordered according to their priority, i.e. for decreasing weights. As expected, the *Cost Reduction* perspective is considered the most crucial for the business immediately followed by *Market Driven Orientation* and *"Brand Attractiveness"*) which is an important aspect especially for demand-driven supply chains, as fashion industry's ones. *Environmental sustainability* instead, although is receiving increasing attention from researchers is still considered the less important by the interviewed managers;
- Tables 2 and 3 show the ABC analysis for the risk factors according to their weights.

In Table 2 we can see that most of the "A" category risk factors are contained in clusters 4 (*"Informative flow"*) and 6 (*"Distribution"*), while "C" risk factors are mainly concentrated in clusters 1 (*"Competitive Environment"*), 3 (*"Offer"*) and 7 (*"Knowledge of customers"*). These results confirm that supply chain efficiency, in terms of correct management of both material (i.e. distribution of fashion products to

Table 2 Distribution of risk factors over the ABC classes for each cluster

Category		A (%)	B (%)	C (%)
Cluster				
1	*Competitive Environment*	0	22	78
2	*Relationship between SC actors*	47	21	32
3	*Offer*	10	10	80
4	*Informative flow*	75	13	13
5	*Process and timing control*	15	38	46
6	*Distribution*	64	27	9
7	*Knowledge of customers*	20	10	70
8	*Quality and Quantity control*	42	42	17

Table 3 Distribution of risk factors over the ABC classes for the time phases

	# factors	A (%)	B (%)	C (%)
Pre-season	78	35	23	42
In-season	6	33	33	33
Post-season	8	38	25	38

final consumers) and informative flow, needs most of the company's efforts for avoiding any possible disruption or delay. On the contrary, competition and the appropriate knowledge and management of the market are not considered highly risky fields, although fashion market trend and customer's tastes are always changing.
Table 3 instead, highlights that, despite the 85% of all the risk factors is related to the Pre-Season phase, they are almost equally distributed over the ABC classes, indicating that only the 35% of them is considered highly crucial for the company.

In brief, the results of the ANP analysis show that the constant research for *Cost Reduction* and, in particular, for the reduction of the *Time to Market* is considered by the interviewed managers as the most crucial aspect for risk management but it is not specific for fashion retailing but is shared by all industries. *Market driven orientation* is, instead, a specific issue for any demand-driven supply chain as those of the fashion industry. In this context, the next section focuses on this aspect and proposes a framework for the optimisation of performances. The proposed model is based on the analysis, during the sales season, of real-time demand data allowing companies to adapt their operations plans according to real market demand.

3 Framework for Optimisation of Supply Chain Performances

After the identification of the main risks, a new SC framework is proposed in order to better manage the most critical areas, i.e. *Time to Market* and *Informative Flow*. This section, then, presents a framework which defines all the processes, material and informative flows that are characteristics of a fashion company operating with a dense network of direct-operated or franchising mono-brand stores. Main purpose of this model (Fig. 1) is to optimise Supply Chain performances through a responsive approach which, during the sales season, analyses actual market demand and adjust operations plans according to it. In order to compare different scenarios, a set of technical and economic Key Performance Indicators (KPIs) is defined.

The process starts from the development of the *New Collection* (A) by the styling office and the definition of the *Demand Forecasts* (C). While the *New Collection* is considered as a simple input for the framework, forecasting is one of the pillars on which all further planning activities are based. In the F & A Industry this process is crucial and particularly complex due to high volatility and unpredictability of demand and is based on historical sales data and characteristics of the new collection

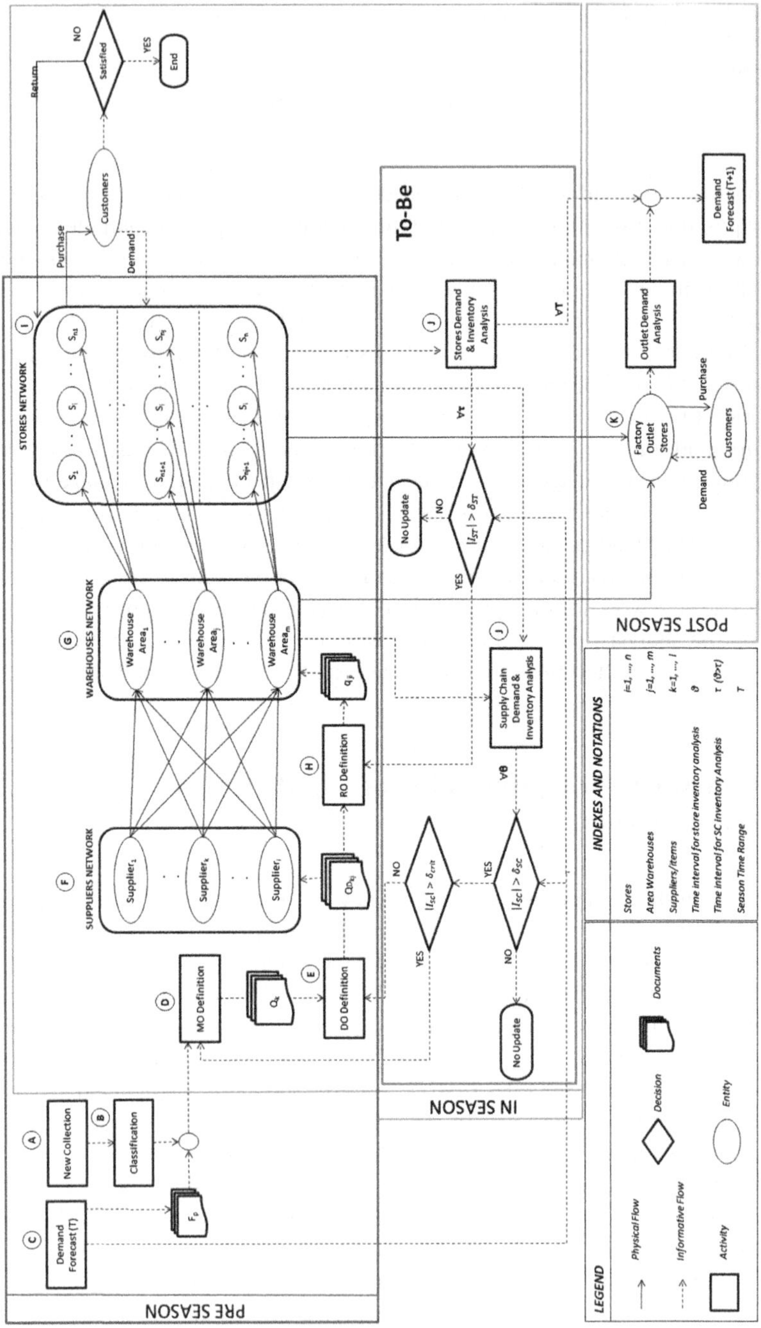

Fig. 1 Framework for the optimization of the supply chain in the fashion retail industry

(*Classification*-B) and stores. Next step is the drafting of *Merchandise Orders* (D), which define purchasing quantities for each item, and *Delivery Orders* (E), which define time and place for products deliveries from *Suppliers* (F). For simplicity, we suppose that the k-th supplier produces the k-th item and delivers it all to the area warehouses in quantity Q_{kj}. The supply process ends with the delivery of goods to the *Area Warehouses* (G) according to the Delivery Orders. At this point, warehouse staff has the task of preparing personalized kits of items to send to the *Stores* (I) according to the *Replenishment Orders* (H). The j-th warehouse supplies only a specific set of n_j stores pertaining to its area. The process described so far defines the material and informative flow that characterizes the Pre-Season phase that, as the name implies, is performed before the beginning of the sales season. The In-Season phase, instead, starts with the first sales recorded in the stores. We suppose that both deliveries form suppliers and replenishments to stores are also performed during the selling season even if they are scheduled before it. Thank to sales data recorded in the stores, it is possible to assess deviations between real sales and forecasts. This analysis represents the core of the proposed approach (To-Be) and defines in real-time how much the demand was under-estimated or over-estimated. If this deviation is higher than a fixed threshold, the model will update all the *Merchandise Orders* and possibly cancel some orders or issue new ones; otherwise it will simply update *Replenishment Orders*, increasing or reducing quantities to be delivered to stores. This adjusting procedure is called *Update Process*. The In-Season phase ends with the collection of unsold goods from stores and warehouses. These unsold items will be then delivered to *Factory Outlet* stores (K) and disposed during the following seasons (Post-Season). It is clear that each item will have a fall in price in proportion to the time of storage in the outlet store, which results in a reduction of the contribution margin. In addition, during this phase, all sales data recorded will be collected and used to draft demand forecasts for the following season (T + 1).

3.1 Key Performance Indicators

After the definition of the new SC model, a set of Key Performance Indicators (KPI) was selected in order to evaluate its performance. This KPI set was defined accordingly to the mainly used process performance measures and according to Lanzilotto et al. (2015), considering both technical and economic aspects.

(a) ***Technical KPIs***

- *Service Level*: It is usually defined as the ratio between orders fulfilled and total orders received. In this context, it is expressed as the ratio between actuals sales recorded and demand received;
- *Forecasting Accuracy*: is defined as one minus the percentage of forecasting error compared to actual sales;
- *Inventory Turnover*: it represents the number of times that inventory is used in a fixed time period and is expressed by the ratio between quantity outgoing the

warehouse and average stock. This indicator is evaluated both for the central warehouse (where outgoing quantities are the item delivered to stores) and for stores' internal warehouses (where outgoing quantities are the sales);
- *Sales Percentage*: it is defined as the ratio between actual sales and quantities delivered to each store. This indicator is specifically defined for the application in the fashion industry. Given the impulsive purchasing behavior of customers, in fact, sales will increase with the availability of product in stores, then it is meaningless to evaluate the pure data on actual sales;
- *Availability*: it is a measure of the event whereby the requested item is not available in the store and is strictly connected to the previous KPI;
- *Out of Stock*: this parameter is strictly connected to the availability and to the service level. It is the define as the number of orders that can not be fulfilled;

(b) ***Economic KPIs***

- *Purchase Cost*: related to the quantity bought before the sales season;
- *Primary Transport Cost*: it refers to the deliveries from the suppliers to the warehouses and is proportional to distances and quantities delivered;
- *Secondary Transport Cost*: related to the replenishment process, i.e. deliveries from warehouses to the stores. As the previous cost item, it is proportional to distances and quantities delivered;
- *Warehouse Management Cost*: it includes fixed and variable costs for the holding of clothing items in the warehouse;
- *Stores Management Costs*: it includes fixed and variable management costs for main and factory outlet stores and for the holding of products;
- *Out of Stock Cost*: it refers to the economic loss connected to the image damage, lost sale and/or lost customer;
- *Revenues*: they are time depending since products suffer a depreciation according to the time of permanence in the store.

3.2 Analysis of the Results for a Traditional Supply Chain

In order to validate Supply Chain performances of the proposed model we used a real case of an Italian Fashion Company which works in the national territory with hundreds of franchising and direct operated mono-brand stores and just a single central warehouse. The data collected from the above-mentioned company concern characteristics of 10 selected clothing items and 10 selected stores, and historical sales data collected over a time range of 6 months (24 weeks) corresponding to the whole Fall/Winter season. We suppose that demand forecasts are equal to the historical sales data. Six different scenarios were defined and simulated by varying the demand profile:

Scenario 1: the demand follows a gaussian distribution with mean equal to the demand forecast and standard deviation equal to 1% of the mean;

Scenario 2: it is equal to the previous case with a higher standard deviation: 10% of the mean value;

Scenario 3: from week 10 to 17 the demand has a peak (3 times greater than forecasts) while during the other weeks it follows Scenario 1;

Scenario 4: from week 10 onwards, the demand has a growing trend equal to 1% each week;

Scenario 5: from week 10 onwards, the demand has a decreasing trend equal to 1% each week;

Scenario 6: from week 10 onwards, stores 1 to 5 have a growing trend—as Scenario 4—while store 6 to 10 have a decreasing trend—as Scenario 5.

The most important result emerged from the analysis of the KPIs are:

- from the economic perspective, the proposed model (To-Be) guarantees better performances. All the cost and revenues items were evaluated together through the *Profit*, which is always higher than the As-Is case. In particular it is more than 5 times greater in Scenario 2, that represents an unexpected peak in demand (Fig. 2a). This implies that the proposed model is able to better follow demand variations during the sales season. This behaviour is also confirmed by the *Service Level* (Fig. 2b), that is never lower than the As-Is case;
- thank to the adjusting process, also the *Forecasting Accuracy* (Fig. 2c) results slightly higher in the proposed model, since it is able to follow market changes.

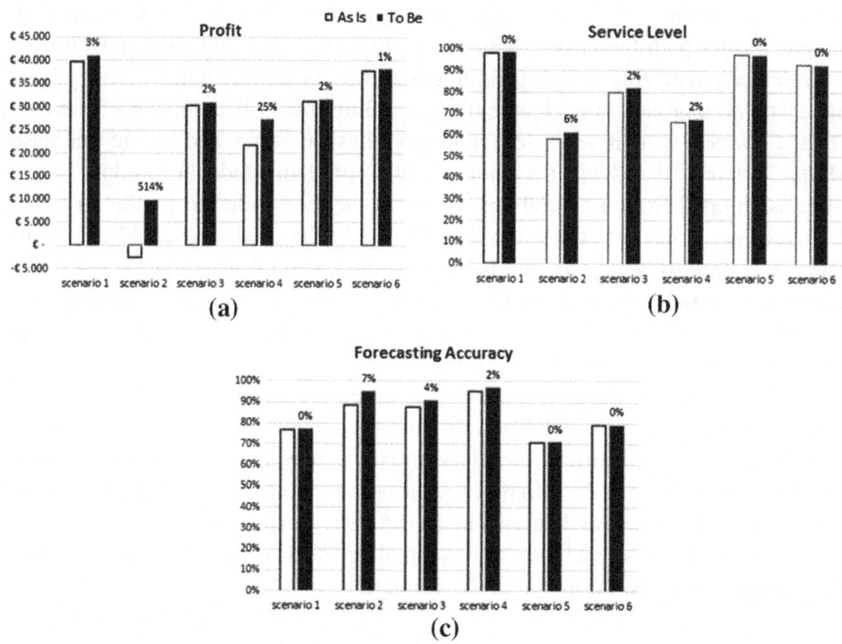

Fig. 2 Comparison of the results for profit and service level

4 The Omni-Channel Strategy: Impact Analysis on a Traditional Supply Chain

The last decades have been characterized by the wide spread of e-commerce and mobile channel purchasing that have deeply changed retail business and management strategies. According to Keller et al. (2014), in fact, Global apparel and footwear market has grown 5% from 2002 to 2015, but online channel have grown 3-4 time faster. Then, in the recent years, one of the main challenge to meet customers' needs is the integration of traditional stores with mobile channels in a new synchronized operating model called omni-channel retailing (Lanzilotto et al. 2014. The diffusion of ICT based tools of retail supply chain has contributed to modify retail operations (Elia and Gnoni 2013). Traditional brick and mortar companies have, in fact, attempted to increase sales and improve profitability by adding online retail channels for consumers (Bretthauer et al. 2010). On the other side, "pure-play" Internet retailers are also opening physical stores or cooperating with traditional retailers (Agatz et al. 2008). It is clear that the simultaneous and integrated management of both physical and mobile channels is not simple and assumes that the supply chain meets the requirements of visibility, accuracy and control of information, flexibility and efficiency. Furthermore, some processes such as inventory management and logistics become extremely critical by adopting an omni-channel retailing strategy. In this context, this section analyses the impact that this new integrated strategy may have on the performances of the traditional Fashion SC. The analysis is focused on the "buy online—pick up in store" strategy, also called "Click & Collect". In this purchasing path the customer buys the product online and then picks it up in a selected physical store or in a pick-up point, thus cancelling home delivery costs. Management approach may change if the retailing model is completely integrated (*Omni-Channel*) or if it separately manages physical stores and online purchases (*Multi-Channel*). If a *Multi-Channel* model is implemented, the product bought online is shipped from a central warehouse to the store selected by the customer; no control is carried out at the store level to verify the product availability. On the contrary, if *Omni-Channel* model is implemented, central warehouses and stores' inventories are synchronized in a centralized inventory. In this case, cornerstone of the system is the *Virtual Inventory System*: it contains inventory data from central warehouses and stores updated in real time.

The framework on which this analysis is based is represented by the As-Is case shown in Fig. 1, with the same case study used in the previous section. By progressively increasing the historical sales data, and the consequently the demand, coming from the online channel, 11 different scenarios were simulated. Table 4 reports the percentage of the historical data used in the different scenarios for the online channel and the physical stores. Each scenario is simulated both with a multi-channel and a omni-channel strategy, for a total of 22 simulation.

Table 4 Simulated Scenarios

Forecast	Scenario										
	0 (%)	1 (%)	2 (%)	3 (%)	4 (%)	5 (%)	6 (%)	7 (%)	8 (%)	9 (%)	10 (%)
Online	0	10	20	30	40	50	60	70	80	90	100
Physical	100%										

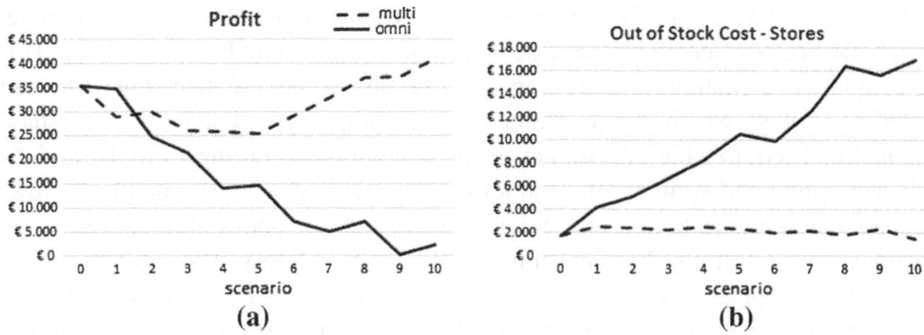

Fig. 3 Profit and out of stock costs

4.1 *Analysis of the Results for the Omni-Channel Strategy*

The most important results obtained from the simulation are shown in Fig. 3. The graphs show the unitary *Profit* and *Out of Stock Cost*, i.e. compared to the total quantity purchased. In terms of Profit (Fig. 3a), compared to a traditional Supply Chain (Scenario 0), the multi-channel strategy is increasingly better when online purchases grow. The opposite situation occurs for the omni-channel strategy. The factor that mostly causes this difference is the Out of Stock Cost. Figure 3b, in fact, shows the Out of Stock Cost and reflects that the replenishments are not optimised for the omni-channel strategy Martino et al. (2016). Stocks are, in fact, consumed by online purchases before they are appropriately restored.

The reduction of the Out of Stock may be obtained by optimising the replenishment policy, i.e. by varying replenishment plans, in terms of frequencies and quantities. Risk analysis (Sect. 2) already identified the replenishment as one of the most crucial processes for a Fashion Retail Industry and the introduction of the Omni-Channel retailing confirms that first result.

5 Conclusions

In the complex and dynamic context of the Fashion & Apparel industry, this paper presents different analysis and models that, integrated in an efficient Decision Support System, may allow companies to optimise performances of their Supply Chains.

After a first risk analysis of the overall structure of a traditional Supply Chain operating in this sector, it emerged that the most critical target is the correct Time Management—i.e. the ability of being responsive to market fluctuations. This paper, then, proposes a model for the optimisation of Supply Chain performances through an in-season deviation analysis and an adjusting process. This model shows an improvement, especially from an economic perspective, in all the simulated scenarios and in particular in the case where an unexpected peak in demand occurs. Nevertheless, this framework focuses on a brick-and-mortar only Supply Chain model which has been overcome by the wide spread of online and mobile purchasing. This "e-commerce revolution" forced companies to evolve in the so-called Omni-Channel Retailing strategy which implies management of both physical and online purchase channel and requires the adjustment of the operations according to it. To include this new integrated strategy, the framework was revised and extended and shows that for the appropriate and convenient management of both online and physical channel, a well optimised replenishment policy is mandatory. Then, the presented multi-method approach can represent a useful system for supporting companies in the strategic decision making process for the definition of the purchasing quantities and operations plans well ahead of the sales season.

References

Agatz NA, Fleishmann M, van Nunen JA (2008) E-fulfillment and multi-channel distribution—a review. Eur J Oper Res 187:339–356

Bretthauer KM, Mahar S, Venakataramanan MA (2010) Inventory and distribution strategies for retail/e-tail organizations. Comput Ind Eng 58:119–132

Brown MJ, Farmer DJ, Ganenthiran N (2013) Recasting the retail store in today's omnichannel world

Elia V, Gnoni MG (2013) Pervasiveness of RFID technology: a survey based on case studies analysis. Int J RF Technol Res Appl 5(1–2):41–61

Hansson M (2011) What impact has a fast fashion strategy on fashion companies supply chain management?

Iannone R, Ingenito A, Martino G, Miranda S, Pepe C, Riemma S (2013) Merchandise and replenishment planning optimization for fashion retail. Int J Eng Bus Manage 5

Iannone R, Martino G, Miranda S, Riemma S (2015) Modeling fashion retail supply chain through causal loop diagram. In: 15th IFAC symposium on information control problems in manufacturing, Ottawa, Canada, vol 28. pp 1290–1295

Keller C, Magnus K-H, Hedrich S, Nava P, Tochtermann T (2014) Succeeding in tomorrows globalfashion market, September 2014

Lanzilotto A, Martino G, Gnoni MG, Iannone R (2014) Impact analysis of a cross-channel strategy in the fashion retail industry: a conceptual framework. In: Proceedings of the summer school Francesco Turco, Senigallia, Italy, pp 58–63

Lanzilotto A, Martino G, Gnoni MG, Iannone R (2015) Impact analysis of a cross-channel retailing system in the fashion industry by a simulation approach. In: 14th international conference on modeling and applied simulation, Bergeggi, Italy, pp 79–88

Martino G, Fera M, Iannone R, Sarno D, Miranda S (2015) Risk identification map for a fashion retail supply chain. In: Proceedings of the summer school Francesco Turco, Naples, Italy, pp 79–87

Martino G, Packianather MS, Yuce B, Iannone R (2016) Optimisation of the replenishment problem in the fashion retail industry using tabu-bees algorithm. In: IFAC-PapersOnLine, Troyes, France, vol 49(12). pp 1685–1690
Rosenblum P (2015) Fast fashion has completely disrupted apparel retail
Saaty TL, Vargas L (2006) The analytic network process. Decision making with the analytic network process, pp 1–26
Sheng L (2014) U.S. fashion industry benchmarking study, June 2014
Shoenbachler DD, Gordon GL (2002) Multi-channel shopping: understanding what drives channel choice
Motorola Solutions (2012) How brick-and-mortar retailers can win the omnichannel battle

Testing and Deploying an RFID-Based Real-Time Locating System at a Fashion Retailer: A Case Study

Antonio Rizzi and Giovanni Romagnoli

Abstract In this paper, the testing and deployment of an RFID-based Real-Time Locating System (RTLS) at a fashion retailer is presented. An RTLS is a combination of hardware and software systems to determine automatically and in real-time the coordinates of an object. Although this goal might be achieved with different technologies, several studies suggest the centrality of RFID as a locating system technology. Nonetheless, the literature presents a lack of studies that investigate on real applications of RTLS in retailing. To fill this gap, our study reports the testing and deploying phases of an RFID RTLS at the Diffusione Tessile store in Pomezia, Rome (Italy). The selected store is, at present, the biggest of the whole firm in terms of number of garments exposed. During a recent refurbishment of the store, the company installed a RTLS composed of 254 antennas linked through multiplexers to 13 readers. The RTLS was designed after an extensive lab-testing phase, and it is provided with two different algorithms for locating garments on the sales floor area. A first installation of the system was finished in early 2015, but this first implementation resulted in some discrepancies amongst different antennas and some corrections were made. In autumn 2015, the system was then tested for the first time in the field: the results of all the testing phase are reported in the paper. This case study sparks interest and suggests several ideas for a deployment of RTLS in a fashion store. Also, and maybe more important, the points of strength and weakness of our implementation could help practitioners and researchers to maximise the benefits of future RTLS implementations in the fashion industry and discover new prospective research topics within this sector.

Keywords RFID · Fashion retail · Real-Time locating system · Deployment · Case study

---

A. Rizzi · G. Romagnoli (✉)
Department of Industrial Engineering, University of Parma,
V.G.P. Usberti 181/A, 43124 Parma, Italy
e-mail: giovanni.romagnoli@unipr.it

A. Rizzi
e-mail: antonio.rizzi@unipr.it

R. Rinaldi and R. Bandinelli (eds.), *Business Models and ICT Technologies for the Fashion Supply Chain*, Lecture Notes in Electrical Engineering 413,
DOI 10.1007/978-3-319-48511-9_17

1 Introduction

The fashion and apparel sector is very dynamic and interesting, and its supply chain (SC) comprehends several actors and various flows of products and information (Kay 2011). The fashion industry, in fact, is characterized by unpredictable demand, high variety and short life cycles of products (Bottani et al. 2016). Garments are somehow similar to "perishable" goods, i.e. when trends change, products become obsolete and their prices decrease significantly. Furthermore, the fashion and apparel SC is often featured by long flow times. The overall turnover of the fashion and apparel SC in Italy in 2015 slightly exceeded 62 billion of Euros, and increased of 1.4% in comparison with 2014 (Crivelli and Bottelli 2016). Under these circumstances, the technology for Radio Frequency Identification (RFID) has already proven to be an effective tool for achieving significant benefits along the entire SC, because it enables to track and trace garments at item level. In the last 10 years, several fashion and apparel manufacturers and retailers started to test and deploy RFID in upstream logistics processes (Madhani 2011) and, more recently, the technology was also tested and deployed at store level, with an ever increasing pace (RFID in retail and apparel 2015 report 2015).

RFID technology, in fact, enables an optimal management of the fashion and apparel SC, and thus it increases the reliability of sales forecasts: fully automated processes are, in fact, more accurate, and the information from the field can be shared in real-time along the whole SC (Kay and Bogart 2005). Inaccurate sales forecasts are particularly undesirable in the apparel sector, due to the nature of its production logic, because apparel is made to stock, and to the typically high margin between production costs and selling prices. In addition to classic RFID processes, such as receiving, replenishment and shipping, the apparel sector is puzzled in the last few years by the possibilities of an innovative system for inventory count and real-time location of items, the so-called Real-Time Locating Systems (RTLS). The RTLS consists in an improvement of a common RFID system with passive tags; however, the main advantages of this solution are automatic real-time inventory counts and real-time location of garments within the sales floor (or backroom) area. The latter, it goes without saying, is an important and high-value-added information that was never delivered by classic RFID systems.

The aim of this study to test and deploy an RFID RTLS by a fashion and apparel retailer in Central Italy, namely Diffusione Tessile (DITE). In particular, we propose new industry-oriented performance indicators to monitor inventory and location accuracy of a RTLS, so as to evaluate its performance and understand how the system can be improved. Furthermore, we believe that our study could sparks interest and suggests several ideas for a deployment of RTLS in a fashion store.

The remainder of the paper is organised as follows: in Sect. 2 we report a brief review of the literature on RFID RTLS. Afterwards, in Sect. 3 we describe our case study, and report the details and results of the testing phases in Sect. 4. Finally, Sect. 5 draws the conclusions of the study.

2 A Brief Review of the Literature on RFID RTLS

According to Li et al. (2016), a RTLS can be defined as a combination of both hardware and software systems capable of determining automatically and in real-time the coordinates of an object within an instrumented area. This goal can be achieved with different technologies, such as:

i. Infrared (Want et al. 1992);
ii. WLAN—Wireless Local Area Networks (Bahl and Padmanabhan 2000);
iii. GPS—Global Positioning System (Song and Eldin 2012);
iv. Ultrasonic (Priyantha et al. 2000);
v. UWB—Ultra Wideband (Ingram et al. 2004);
vi. Vision analysis (Gu et al. 2009);
vii. RFID (Ni and Patil 2003).

For considerations on and comparisons of different RTLS technologies, we refer to (Gu et al. 2009; Li et al. 2016; Liu et al. 2007). However, although RFID is neither the most accurate nor the most convenient solution, we note that several papers suggest its centrality as a locating system technology (Bouet and Dos Santos 2008; Gleser and Ondráček 2014; Liu et al. 2015; Ni and Patil 2003). RFID is a wireless communication technology that uses electromagnetic fields to identify, track and trace RF compatible tags (e.g. integrated circuits) attached to objects. The RFID system comprehends tags, readers and antennas, and it is typically used in complex indoor environments such as warehouses, factories, hospitals and offices. RFID tags can be active or passive, differing in the method of powering the tags and, consequently, in communication ranges. Active tags use internal sources of power to transmit their ID signals and stored information, while passive tags have no internal power sources and reflect RF signals transmitted from readers. Several analysts (e.g. Technavio 2014; Markets and Markets 2015) predict a rapid growth of active RTLS and, therefore, a positive market trend can be assumed, although real-time locating of mobile objects based on active RFID remains far from being ubiquitous. Passive-RFID systems, instead, are characterized by costlier infrastructures, as the reading distance is more limited than for active solutions; nonetheless, these systems offer a huge advantage on tags' costs, and this advantage becomes even more important in applications with several thousands of tagged objects. Active locating solutions, in fact, have seen only limited deployments: for applications where many objects need to be tagged, RTLS based on cheap passive tags seems economically more feasible. Still, the reading range and the accuracy in passive RTLS must also be considered.

Theoretically, a passive-RFID RTLS could work in all classical RFID-frequencies; however, the limited activation-/reading-distance of Low Frequency- and High Frequency-tags does not seem useful for applications in retail or logistics. The Ultra High Frequency (UHF) range, with a reading range of 3 m and more, is more promising for locating applications in the cited domains, especially as UHF-tags for identification are being more and more deployed in these domains. In the last two decades, in fact, RFID technology has been deployed in a variety of sectors, from construction (Song et al. 2006) to logistics (Oztaysi et al. 2007), and from fast moving consumer goods (Bertolini et al. 2015) to fashion and apparel (Rizzi et al. 2016). The possibility of using RFID technology for locating tags, and therefore the items the tags are attached to, is more than 10 years old. Amongst the first studies to propose similar solutions there are the ones from Hightower et al. (2001) and from Ni and Patil (2003). According to Gleser and Ondráček (2014), the market of RTLS based on RFID is at the beginning, with high demand and recognized potentials. Also, real-time locations with RFID tags has seen several important contributions (see for example Zhang et al. 2007). In order to achieve high RTLS accuracies, different principles can be used and combined, such as (i) Time of Arrival (ToA); (ii) Roundtrip Time of Flight (RToF); (iii) Time Difference of Arrival (TDoA); (iv) Angle of Arrival (AoA); (v) Received Signal Strength (RSS); (vi) Phase of Arrival (PoA); (vii) identification and (viii) movement forces (Liu et al. 2007; Gleser and Ondráček 2014).

However, scientific literature generally agrees on the fact that RTLS RFID systems must always be adapted on the individual cases and local conditions in which it will be running, and a future trend shall be how to deploy sensors to improve positioning accuracies in real environments. To this aim, there is a general lack of studies that investigate on real applications of RTLS in retailing. To fill this gap, our study reports the testing and deploying phases of an RFID RTLS at the DITE store in Pomezia, Rome (Italy).

3 The Case Study: RTLS at DITE, Pomezia

Diffusione Tessile is part of one of the largest groups in the world for high quality women's clothing. The company owns and manages a set of factory outlets and sells inventories of textiles, apparel and fashion products from previous seasons. The company has a main Distribution Centre (DC) in northern Italy and 20 stores in Italy and across Europe. DITE has performed the first RFID pilot in 2012. Currently, RFID in DITE involves 4 stores, with the plan of a full deployment (all products, all stores) in the next 5 years. Items are tagged with RFID chips at the DC, and these tags move along with garments throughout the whole supply chain, up to the checkout counters, where cashiers complete the sale process and remove tags from garments.

In late 2014, DITE planned to test and deploy a RTLS in its Pomezia store (Rome, Italy). The driving causes of the implementation of a RTLS at this store were:

i. enable automatic inventory counts;
ii. helping customers to locate the items they were looking for.

Due to the size of the store (over 1,800 m^2, that makes it the largest DITE store in terms of size and number of items exposed), in fact, the company deemed that up to a 10% of the total number of sales could be lost due to the impossibility of customers to find the precise products they were looking for, although they were available at the store. Furthermore, the size of the store makes it difficult to perform frequent inventory counts via handheld readers, for that operation takes on average between 1 and 2 h of a skilled associate.

To avoid this, the company installed during a recent refurbishment of the store (2013) 254 antennas linked via multiplexers to 13 readers. The system integrator for the RTLS was ID Solutions SRL, a spin-off company of the University of Parma, with expertise in automatic acquisition and processing of data to generate value added indicators. A first installation of the system was finished in early 2015, but this first implementation resulted in some discrepancies amongst different antennas: while some antenna read several thousands of items per day, others saw only few garments on average, and this suggested that these antennas were misplaced. After an accurate control, in fact, some antennas were found to be badly positioned and moved to better reading positions. In late 2015, the system was tested for the first time in the field. The following sections report both the lab and the on field testing of the RTLS system inside Pomezia store.

4 Testing the RTLS at DITE

4.1 Lab Testing

During lab testing, two different algorithms were developed for locating garments on the sales floor area. According with the company, the aim was that of developing easy-to-implement solutions, to reduce the time-to-market of the first RTLS prototype. The first algorithm is based on the number of reads of any given tag during a certain period, and it relies on Eq. (1).

$$X_i = \frac{\sum_{a=1}^{n}(X_a n_a)}{\sum_{a=1}^{n} n_a} \quad Y_i = \frac{\sum_{a=1}^{n}(Y_a n_a)}{\sum_{a=1}^{n} n_a} \tag{1}$$

Where:

- X_i (Y_i) indicates the x-coordinate (y-coordinate) of a given tag calculated by the RTLS;
- X_a (Y_a) indicates the x-coordinate (y-coordinate) of any antenna who read the tag at least once during the period (note that the x- and y-coordinates of all antennas are known from the installation map of the RTLS);
- n_a stands for the number of times the given tag has been read from antenna a during the time period.

It is clear from (1) that the higher the number of reads of any given tag by antenna a, the closer to a the tag will be expected. A different approach is used for the second algorithm: in this case, in fact, it is not the number of reads but the Received Signal Strength Indicator (RSSI) that drives the localization of the tag. The RSSI value, in fact, is negative and, typically, it increases in absolute value when the distance between the antenna and the tag increases. Therefore, by using the reciprocal of the absolute value of RSSI (practically always positive and increasing when the distance between the antenna and the tag increase) the locating algorithm will work according to the equations in (2). In those equation, X_a and Y_a indicate, respectively, the x- and the y-coordinate of any antenna who read the given tag at least once, while $RSSI_a$ indicates the strength of the signal between the given tag and antenna a. We tested both these algorithms in the lab and they performed well: therefore, we implemented both these algorithms in the RTLS software, to test them both on the field and assess which solution could deliver better performances in Pomezia store.

$$X_i = \frac{\sum_{a=1}^{n} \left(X_a \frac{1}{|RSSI_a|}\right)}{\sum_{a=1}^{n} \frac{1}{|RSSI_a|}} \quad Y_i = \frac{\sum_{a=1}^{n} \left(Y_a \frac{1}{|RSSI_a|}\right)}{\sum_{a=1}^{n} \frac{1}{|RSSI_a|}} \tag{2}$$

4.2 Preliminary on Field Testing

The testing of the system in static condition has started in October 2015: during this preliminary test, we were bound by some manual operations for crosschecking RTLS performances, and therefore we considered only a few hundred items. This testing phase was aimed at a preliminary evaluation of performances of the RTLS operated according to the equations indicated in (1) and (2). We proceeded as follows.

Firstly, according to DITE requests, we parted the Pomezia store in 26 different areas, as indicated in Fig. 1. The definition of different areas makes use of structural elements in the sales floor (cabinets, shelves, racks, counters etc.) and are defined according to the experience of store operators. In particular, we started our tests in 8 different areas (namely, Ala dietro; Accessori; M2; Maglieria; Marina; S1.1; S4.2; S5.2): we performed inventory counts with handheld readers in all these areas. Reading power of handhelds (HHs) were set relatively low (10–30%), to read only tags at a 1/2 m distance. Also, the EPC values resulting from HH inventory count were randomly checked to assess their validity. These values were used as a baseline, i.e. all EPCs read from HH were considered to be exactly where we found them. Then, the EPC values read with HHs were compared with the ones available from the RTLS operated with the algorithms (1) and (2). The comparison could give the following results:

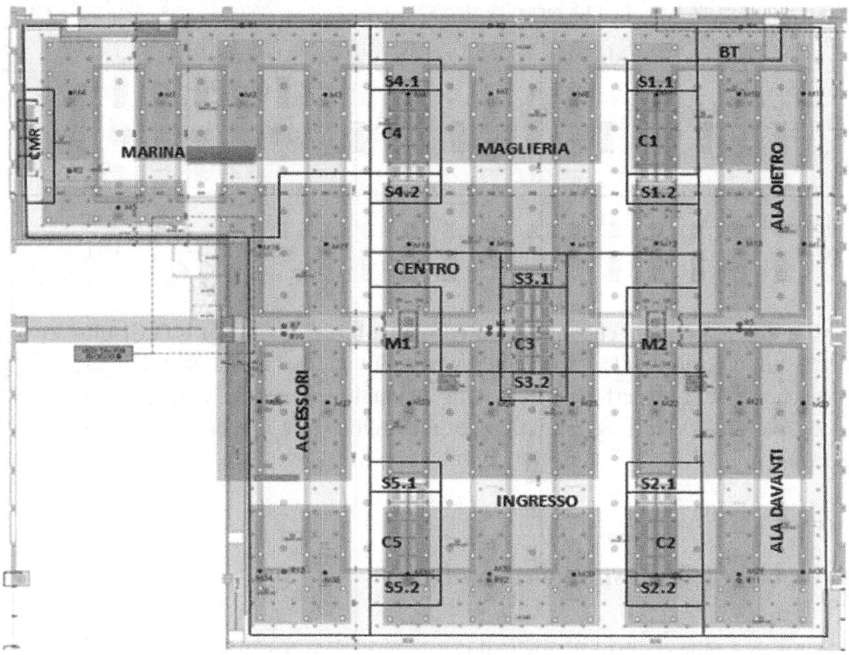

Fig. 1 A map of Pomezia store with its 26 areas for the RTLS

- **Right location**: when the RTLS locates a tag in the same position as the HH;
- **Neighbouring location**: when the RTLS locates a tag in a neighbouring area to the one indicated via HH;
- **Wrong location**: when the RTLS locates a tag in an area different from those reported above (neither exact nor neighbouring);
- **Missing location:** when the RTLS has never read the tag found via HH.

The results of this preliminary testing is indicated in Table 1. The table reports that algorithm (1) performs slightly better than (2): in detail, the RTLS correctly locates 75.7% of items with (1), against 72.5% of (2). Also, the location accuracies increase to 85.5% for (1) if we also consider neighbouring areas, against 85.2% of (2). We noticed, in fact, that garments were more often located in close areas when the areas they were in were small and/or when they were on the border between two different areas, i.e. when they could be found by looking in one of the two adjacent locations. Finally, it is important to note that the wrong locations of items are relatively rare (less than 2% with both location algorithms), and the biggest problem of our preliminary testing was that 13.0% of tagged garments that were never read by RTLS readers.

Table 1 Preliminary testing of RTLS in static conditions

	Algorithm (1)		Algorithm (2)	
Location	N. of items	Percentage	N. of items	Percentage
Correct location	256	75.7	245	72.5
Close location	33	9.8	43	12.7
Wrong location	5	1.5	6	1.8
Missing location	44	13.0	44	13.0
Total	338	100.0	338	100.0

For this reason, we chose to proceed with a broad review of RTLS software and deployed algorithm (1). Therefore, we removed the limits that bounded us to crosscheck RTLS performances with manual operations, and proceeded with extensive on field testing. The protocol for such testing is reported in the following Section.

4.3 On Field Testing Protocol

This testing phase aimed at evaluating technological performances of the RTLS. It was performed in December 2015. The first question we needed to answer is what kind of RTLS performances are more appealing for DITE store. According with Sect. 3, we decided to implement two different performance indicators:

i. Inventory accuracy, i.e. number of items physically located on the sales floor read by RTLS readers;
ii. Location accuracy, i.e. percentages of items located by the RTLS in correct or neighbouring locations.

To do so, however, we needed a control source of information. As in the previous section, we used handheld data, as shown in Fig. 2. To do so, we developed a web-based procedure that crosschecked RTLS data with Handheld data and delivered inventory and location accuracy as an outcome.

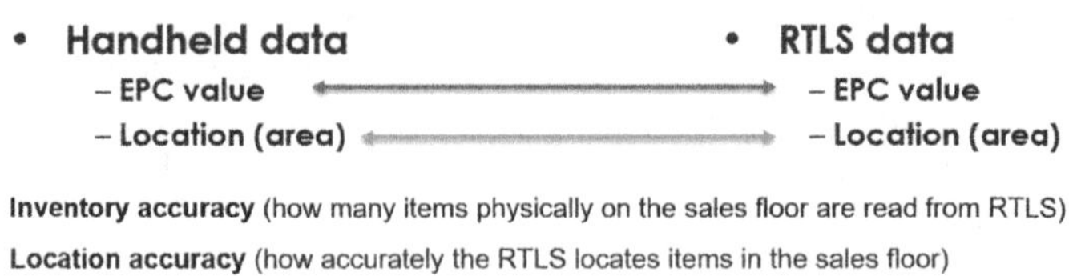

Fig. 2 Comparison between RTLS data and the control sources (HH)

4.4 *Testing Results and Discussion*

The results of the test are reported in Tables 2 and 3. As Table 2 reports, the value of inventory accuracy has greatly increased, with respect to the preliminary testing phase. The number of missing items decreased from 13.0 to 7.0%: this might be due to several factors, such as the reduced size of the preliminary test sample, the longer operating time of the RTLS during the 2015/16 Autumn/Winter season, and the review of the RTLS software performed before the proper testing phase.

Another important factor is the great variety of inventory accuracy values. Given an average inventory accuracy of 93.0%, in fact, area accuracies range from 50.0 to 100.0%. However, this may be explained by considering the great variety of items located in different areas. It is important to note, in fact, that the definition of evenly

Table 2 Inventory accuracies per area of the store

Area	HH reads	RTLS reads	Inventory accuracy (%)
Accessori	2,349	2,179	92.8
Ala Davanti	2,601	2,416	92.9
Ala Dietro	2,524	2,289	90.7
BT	20	16	80.0
C1	0	0	–
C2	6	5	83.3
C3	8	5	62.5
C4	0	0	–
C5	2	2	100.0
Centro	1,199	1,018	84.9
CMR	4	2	50.0
Ingresso	1,471	1,402	95.3
M1	44	27	61.4
M2	56	31	55.4
Maglieria	5,046	4,811	95.3
Marina	4,166	3,926	94.2
S1.1	30	26	86.7
S1.2	41	39	95.1
S2.1	29	27	93.1
S2.2	18	16	88.9
S3.1	25	22	88.0
S3.2	21	20	95.2
S4.1	45	38	84.4
S4.2	29	28	96.6
S5.1	14	13	92.9
S5.2	4	4	100.0
Total	19,752	18,362	93.0

Table 3 Inventory accuracies per area of the store

Area	RTLS reads	Right location (%)	Neighbouring location (%)	Wrong location (%)
Accessori	2,179	93.4	6.3	0.2
Ala Davanti	2,416	95.1	3.2	1.7
Ala Dietro	2,289	98.4	1.4	0.2
BT	16	37.5	50.0	12.5
C1	0	–	–	–
C2	5	80.0	20.0	0.0
C3	5	40.0	60.0	0.0
C4	0	–	–	–
C5	2	0.0	100.0	0.0
Centro	1,018	29.2	69.8	1.0
CMR	2	50.0	50.0	0.0
Ingresso	1,402	93.4	4.9	1.8
M1	27	88.9	7.4	3.7
M2	31	100.0	0.0	0.0
Maglieria	4,811	92.4	7.4	0.2
Marina	3,926	93.0	6.9	0.1
S1.1	26	69.2	30.8	0.0
S1.2	39	64.1	35.9	0.0
S2.1	27	92.6	7.4	0.0
S2.2	16	87.5	12.5	0.0
S3.1	22	77.3	22.7	0.0
S3.2	20	80.0	15.0	5.0
S4.1	38	73.7	26.3	0.0
S4.2	28	82.1	17.9	0.0
S5.1	13	69.2	15.4	15.4
S5.2	4	100.0	0.0	0.0
Weighted average		90.0	9.4	0.6

populated areas of the store could greatly increase the performances of the RTLS. During the test we did not follow this guideline because, for the sake of time, we decided to stick to the already existing store areas, known by store associates and developed for different purposes. However, for the future, it will be interesting to redesign store areas for RTLS purposes, and then pass them to the staff for RTLS operations.

The values of location accuracies are reported in Table 3. As in the preliminary testing, we chose to indicate not only Right and Wrong locations, but also Neighbouring ones. This is due to the fact that store areas are not clearly marked at a physical level, but only indicated on a map (see Fig. 1). Due to this issue, University staff during the test, as well as store associates during daily operations, could mistakenly take one area for another neighbouring one, and therefore look for items also in adjacent store areas. For this reason, garments located in neighbouring areas could also be found by store associates looking for them, and this information is not as wrong as the location of an item in a non-neighbouring location.

Table 3 indicates a weighted average of correctly located items equal to 90.0%, i.e. 9 garments out of ten, when identified by the RTLS, are located in the correct area. Although this value might not be considered encouraging, if we also consider neighbouring locations, the location accuracy increases to 99.4%. Although we cannot prove it with the current test phase, our impressions during the tests was that items located in neighbouring areas were often situated on the or close to the edges between bordering areas. For this reason, the location via handheld readers could, in some cases, be mistakenly considered in an adjacent area, and therefore be a wrong control source of data, instead of an RTLS mistake.

In addition, it is important to note that the location accuracy values are also influenced by the great variability of the population of garments per area: therefore, the same considerations made for inventory accuracy, are also valid in this case. To conclude, the performances of locating accuracy are definitely satisfactory, if neighbouring areas are also considered. A notable exception to this statement is given by the location accuracy of the "Centro" area. Although this area is quite populated, in fact, the RTLS located around 70.0% of the items found there by handheld in neighbouring areas. Unfortunately, we still do not have a decent explanation for this outlying value, and therefore we plan to repeat the test soon.

5 Conclusions

An RTLS is a combination of hardware and software solutions to determine automatically and in real-time the position of an object. Several different technologies are available to this aim, and RFID plays a central role as a RTLS technology. However, scientific literature lacks in industrial applications of RFID RTLS, especially in fast moving retail stores. Our study reports in detail the testing and deploying phases of an RFID RTLS at a major store of a fashion and apparel retailer, namely Diffusione Tessile. In particular, we proposed two industry-oriented performance indicators to monitor performances of the said system, and understand its improvement possibilities.

The results show that the tested system is effective both for automatic inventory counting and real-time locating garments on the shop floor area. To measure these performances, we proposed and explained how to calculate inventory and location accuracies, with respect to a control source of data, e.g. handheld reader data. Our system showed average inventory accuracy of 93.0%, and location accuracies of 90.0%, that reaches the value of 99.4% if we also comprehend neighbouring areas as acceptable locations. It is very important to note that these values were achieved in a store that counts about an average of 20,000 items on the shop floor, and during opening hours, with several hundreds customers within the store.

For these reasons, the paper sets an interesting standard, it sparks interest and suggests several ideas for future deployments of an RTLS in a fashion store. Nonetheless, our study also presents weaknesses, that can be considered as future research trends. They are reported below.

1. The algorithms proposed for real-time location are quite simple and, consequently, rather weak. In our case, the achieved location accuracies were acceptable, as items needed to be allocated in areas with surfaces of, at least, 9 m^2. However, literature presents several studies where RFID RTLS have shown much higher location accuracies.
2. The areas used for locating garments are extremely heterogeneous: this is due to the choice of sticking to the already existing locations, well known to store associates. However, these areas count numbers of garments from a few units to over 5,000, to the detriment of RTLS performances.
3. Some can still be done to improve inventory accuracy: the average value of 93.0% is very good, but perfectible. To this aim, we plan to continue to improve and test the system, until we are fully happy with its performances.
4. The same, at a lesser extent, applies to location accuracy. The 90.0% of correctly located items could be increased, either by marking more clearly different locations within the store, at least during tests, or by improving the location algorithms.
5. Another important factor we did not take into account is the time lag between the last known RTLS read of the tag and the time of handheld reads. This could help to improve our understanding of the RFID RTLS.
6. Similarly, understanding how the RTLS behaves with moving garments could greatly help to understand the issues at hand.
7. Also, it may be possible to find correlations either between density of garments and inventory accuracy in given areas, or between classes of products with peculiar characteristics (e.g. jackets, trousers, t-shirts etc.) and inventory accuracy.
8. Finally, an economic evaluation of the increase in sales after the implementation of the RTLS system could suggest directly the economic feasibility of RFID RTLS in the fashion and apparel retail sector. At present, we are using the RTLS (i) to monitor how much every store area contributes to the overall profit of the store, and (ii) to understand how important is the contribution in terms of sales of the garments located in display areas inside the store, close to the changing cabinets.

At present, the authors are working on some of these topics for upcoming research.

References

Bahl P, Padmanabhan VN (2000) RADAR: an in-building RF-based user location and tracking system. In: INFOCOM 2000. Proceedings of the nineteenth annual joint conference of the IEEE computer and communications societies, pp 775–784

Bertolini M, Bottani E, Romagnoli G, Vignali G (2015) The impact of RFID technologies on inventory accuracy in the apparel retailing: evidence from the field. Int J RF Technol Res Appl 6(4):225–246. doi:10.3233/RFT-150069

Bottani E, Montanari R, Romagnoli G (2016) Improving sales turnover in fashion retailing by means of an RFID-based replenishment policy. Int J RF Technol Res Appl 7(1):65–86. doi: 10.3233/RFT-150072

Bouet M, Dos Santos AL (2008) RFID tags: positioning principles and localization techniques. In: 2008 1st IFIP Wireless Days, WD 2008, pp 1–5. doi:10.1109/WD.2008.4812905

Crivelli G, Bottelli P (2016) I big della moda fanno sistema per crescere, consultato il 29/02/2016, reperibile a: http://www.ilsole24ore.com/art/notizie/2016–02-25/i-big-moda-fanno-sistema-crescere–073233.shtml?uuid=ACaiDibC. Il Sole 24 Ore. Milan (Italy). Retrieved from http://www.ilsole24ore.com/art/notizie/2016-02-25/i-big-moda-fanno-sistema-crescere–073233.shtml?uuid=ACaiDibC

Gleser A, Ondráček O (2014) Real time locating with RFID: comparison of different approaches. In: 2014 24th International Conference Radioelektronika, RADIOELEKTRONIKA 2014—Proceedings, pp 5–8. doi:10.1109/Radioelek.2014.6828470

Gu Y, Lo A, Niemegeers I (2009) A survey of indoor positioning systems for wireless personal networks. IEEE Commun Surv Tut 11(1):13–32. doi:10.1109/SURV.2009.090103

Hightower J, Vakili C, Borriello G, Want R (2001) Design and calibration of the spotOn ad-hoc location sensing system. Seattle (WA), USA

Ingram SJ, Harmer D, Quinlan M (2004) Ultra-wide band indoor positioning systems and their use in emergencies. In: Record—IEEE plans, position location and navigation symposium, pp 706–715. doi:10.1109/PLANS.2004.1309063

Kay M (2011) Straight talk about item-level RFID. Appar Mag 51(2):1–7

Kay M, Bogart S (2005) Pulling the trigger on RFID in apparel retail. Appar Mag 47(3):44–46

Li H, Chan G, Wong JKW, Skitmore M (2016) Real-time locating systems applications in construction. Autom Constr 63:37–47. doi:10.1016/j.autcon.2015.12.001

Liu H, Darabi H, Banerjee P, Liu J (2007) Survey of wireless indoor positioning techniques and systems. IEEE Trans Syst Man Cybern 37(6):1067–1080. Retrieved from http://ieeexplore.ieee.org/stamp/stamp.jsp?tp=&arnumber=4343996&tag=1

Liu Y, Han F, Xu X, Zhang H (2015) Measuring the uncertainty of RFID location streams based on optimal estimation particle filter. Int J Distrib Sens Netw 2015(Article Id: 758391):8

Madhani PM (2011) RFID deployment: fast fashion retailing. SCMS J Indian Manage 8(2):40–51

Markets and Markets (2015, Apr.) Real-Time Location Systems Market worth $3.70 Billion by 2020. http://www.marketsandmarkets.com/PressReleases/real-time-location-systems.asp

Ni LM, Patil AP (2003) LANDMARC: indoor location sensing using active RFID. In: Proceedings of the First IEEE international conference on pervasive computing and communications, 2003. (PerCom 2003), pp 407–415. doi:10.1109/PERCOM.2003.1192765

Oztaysi B, Baysan S, Dursun P (2007) A novel approach for economic-justification of RFID technology in courier sector: a real-life case study. In: 2007 1st Annual RFID Eurasia, Istanbul, pp. 1–5. doi:10.1109/RFIDEURASIA.2007.4368147

Priyantha NB, Chakraborty A, Balakrishnan H (2000) The Cricket location-support system. In: Proceedings of the 6th annual international conference on mobile computing and networking—MobiCom '00, pp 32–43. doi:10.1145/345910.345917

RFID in retail and apparel 2015 report (2015) RFID J. Retrieved March 10, 2016, from https://www.rfidjournal.com/purchase-access?type=Article&id=13556&r=%2Farticles%2Fview%3F13556

Rizzi A, Romagnoli G, Thiesse F (2016) A new framework for RFID use cases in fashion and apparel retailing. Int J RF Technol Res Appl 7(2–3):105–129. doi:10.3233/RFT-150075

Song J, Haas CT, Caldas CH (2006) Tracking the location of materials on construction projects. J Constr Eng Manage 132(9):911–918. doi:10.1061/(ASCE)0733-9364(2006)132:9(911)

Song L, Eldin NN (2012) Adaptive real-time tracking and simulation of heavy construction operations for look-ahead scheduling. Autom Constr 27:32–39. doi:10.1016/j.autcon.2012.05.007

Technavio (2014, Dec.) Global Real-time Location System (RTLS) market 2015–2019. http://www.technavio.com/report/global-real-time-location-system-rtls-market-2015-2019

Want R, Hopper A, Falcao V, Gibbons J (1992) The active badge location system. ACM Trans Inf Syst (TOIS) 10(1):91–102

Zhang Y, Amin MG, Kaushik S (2007) Localization and tracking of passive RFID tags based on direction estimation. Int J Antennas Propag 2007:1–9. doi:10.1155/2007/17426

From Financial Merchandise Planning to Supply Chain Design and Execution

Augusto Bianchini and Marco Tricase

Abstract For years agents believed that technology could have bridged the gap in fashion companies. This has not come true. Significant investments aimed at implementing complex systems have often failed. Indeed, they have not been able, in a simple, flexible and comprehensive way, to integrate all the processes that, by definition, are changeable and not only influenced by deterministic factors. Therefore, it proves necessary, in complex organizations, to promote those best practices and habits that support and enhance personal freedom, judgments and hypotheses. This is the process by which the retailer seeks to provide the right amount and quality of the right merchandise in the right store at the right time, while also seeking to meet the financial goals of the company. This project, developed by the collaboration between the Department of Industrial Engineering (University of Bologna) and K.Group, aims to show how financial planning of Merchandise Planning may be implemented in a major Italian Fashion Retail Company, presenting the preliminary plan to integrate it with the specific processes of Supply Chain Planning and Execution, hence highlighting achievements, methodology and technological resources in terms of: data management (normalization and load data), business intelligence (score carding, dashboards, reporting, analysis), predictive analytics (clustering, simulation), performance management (budgeting, planning and forecasting, profitability), workflow management and data integration.

Keywords Merchandise Planning (MP) · Business intelligence · Predictive analytics · Performance management · Data integration

---

A. Bianchini
Department of Industrial Engineering, University of Bologna, Bologna, Italy
e-mail: augusto.bianchini@unibo.it

M. Tricase (✉)
K.GROUP, Perugia, Italy
e-mail: m.tricase@orangecompany.it

R. Rinaldi and R. Bandinelli (eds.), *Business Models and ICT Technologies for the Fashion Supply Chain*, Lecture Notes in Electrical Engineering 413,
DOI 10.1007/978-3-319-48511-9_18

1 Introduction

Current planning systems are *out-of-date* and don't effectively address today's requirements for a comprehensive channel planning environment (Fig. 1). Most retailers believe that their current planning applications are ineffective and not able to support the complex analysis required to optimize decisions and ultimately meet customer demand, in a scalable and a flexible planning solution.

For most retailers, planning is a cyclical endeavour, repeating tasks on a weekly, monthly, quarterly and even annual basis. These cycles are completed multiple times throughout a planning framework, regardless of any given channel. Although integration of people, processes and technology have seen some progress in the last few years, many retailers still need to improve many organizational aspects of the planning process (Sacchi 2015). Technology tools are available in the marketplace to support advanced planning, incorporating customer insights and trends, cross channel integration, competitive information, and real-time data. However, such tools are only a part of the solution that requires coordinated and *up-to-date* processes (BRP 2015).

In the retail fashion industry, planning of merchandise may be described as a systematic approach aimed at maximizing return on investment through sales planning, inventory, space and assortments, increasing profits and minimizing markdown and stock-outs. Merchandise management is the process by which a retailer attempts to offer the right quantity of the right merchandise, in the right place at the right time, complying with a company's financial budget (Kunz 2005) The merchandise management approach shifts the focus of key features of fashion goods: seasonality, limitation, localization of resources and different life cycles require unique policies and procedures. Since merchandise selection is often crucial for the success or failure of the business, the retail buyer has the great responsibility of best locating, securing, pricing and promoting the merchandise assortment (Bini 2011).

The financial budget is the starting point of a Merchandise Planning (MP) process because merchandise cannot be purchased without a financial commitment,

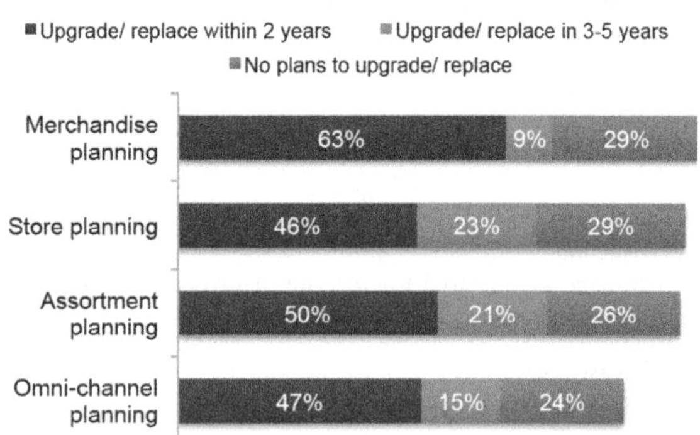

Fig. 1 Application plan (BRP 2015)

although the product research and the assortment development are actually parallel activities (Kunz 2010). Companies need to know what their financial capacity is to support purchases of goods together with the time at which such availability is actual. Hence, top management must allocate specific amounts of money in order to constrain decisions of different divisions on product assortment, that will meet the consumers demand in each specific season within financial boundaries (Iannone et al. 2013).

Once financial availability is known, fashion firms fix their objectives in terms of sales budget (for each season) and category products. When developing a sales budget, a retailer should be able to predict how well product categories are going to sell: from this point of view, previous sales represent the most important source of information. Other useful information regarding future sales may be provided by customer feedback or fashion and trend services (Granger and Sterling 2012).

The budget of sales is either a pre-season and in-season planning (and control) process. During its former phase, the plans are developed within a six-month period before the sales season. Historical data are often used to identify seasonal trends, although a proactive approach to management is crucial. The planner, indeed, should be looking forward in order to timely define promotional activities. On the other hand, in-season planning and control phase involves a management model alert to continuous comparison between current performance and pre-season schedule: as a result, since predictions may be adjusted according to current sales, the planner is able to determine whether stocks are sufficient or not to support the demand. From this point view, readjustment planning is a continuous activity (Retailing Management 2013).

Once sales objectives are defined, fashion companies may also define gross margin objectives for category products to design a merchandise strategy. Therefore, during purchasing phase, buyers would be able to buy goods at a cost that ensures the margin targets (D'Avolio et al. 2015).

Buyers must set the right quantities of goods to be purchased in order to avoid on one hand stock-outs and on the other a surplus of merchandise (in order to minimize inventory costs).

It is therefore clear how a well-structured MP process involves financial, sales, purchasing and assortment product planning in which all business functions take part: it goes without saying that an integrated and fluid framework is necessary to achieve the objectives of the company (La Fleur 2010). The MP process is able to introduce an enough level of automation, that is also related to a high level of workplace safety (Bianchini et al. 2014, 2015).

2 Methodology

The real challenge is thus the realization of a flexible and scalar Merchandise Planning (MP) system fully integrated with existent applications which is able to include directional economic objectives, product development processes, specific

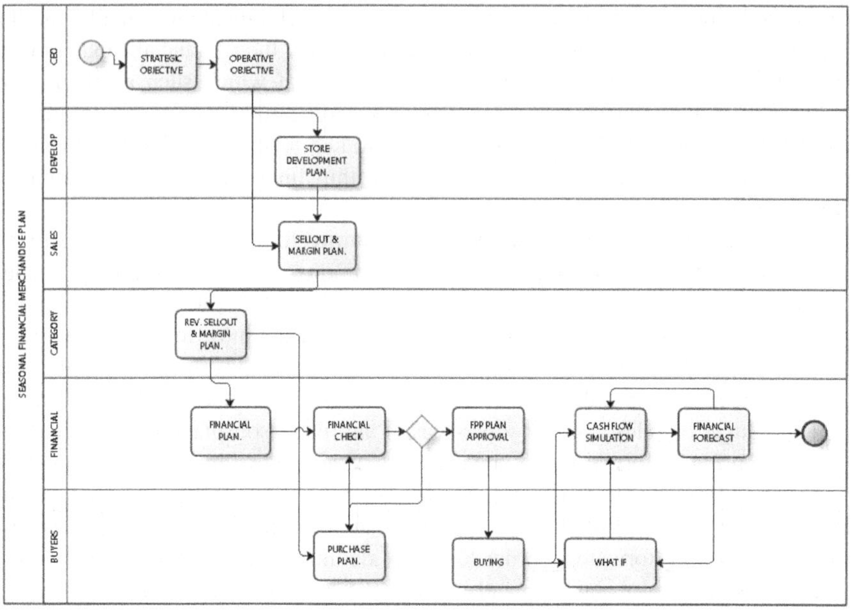

Fig. 2 Financial merchandise planning process

dynamics of the market (network improvement, channel needs and performances, months, store cluster) and operational sectors, in a non-invasive way and pursuant to a company's financial capability.

In this section we will show the methodology (as shown in Fig. 2) applied in a large Italian retail fashion company, with almost 400 stores worldwide (either directly operated stores or franchising), characterized by only one brand on several formats. The main feature of the latter is the Italian fashion approach focused on incessant renewals of good quality products at competitive prices, through seasonal collection (or even flash) strategies.

This pricing constraint involves pursuing issues, particularly tough when dealing with foreign resources, which in turn strongly influence the financial capability of the firm. Thus, the main purpose of our project concerned a stable and retroactively adjustable link between store development planning processes and potential sales development (by article and price point), in such a way as to define the needed stock to be purchased. Such stock, according to different suppliers, delivery period, payment deals and importation issues, gives rise to a purchase budget (together with the monthly financial outflow it implies) that executive managers should approve in order to generate a comprehensive seasonal plan which buyers must respect.

Consequently, buyers' activities, by definition tied to the emotional sphere of the agents, are not damaged by the clear and well-defined targets (references, quantities, costs, product or group mark-up) they must achieve in order to meet directional objectives.

2.1 *Store Development Planning*

Store Development Planning allows the development of an investment plan leading to sales network widening based on the information gathered from new openings, transfers, reorganizations and surface forecasts. According to historical benchmarks, stores are clustered allowing the assignment of variable costs according to their width and fixed costs for each cluster. The result is a P&L account for each store defined up to profits before taxes, thanks to a detailed planning of revenues, cost of goods sold, general, fixed and indirect costs. The above argument holds for any store, even the so called "generic" stores that give rise to a P&L by channel. In any case, annual costs and revenues are translated in monthly units and turned over to sell-out and financial planning models.

2.2 *Sell-Out and Margin Planning*

According to strategical planning of objectives and targets, a company's sales manager fills out the Financial Product Plan (FPP) based on information coming from both store development and goods categories sell-out in previous seasons. Store clustering allows a quick forecasting of future effects of several parameters such as mark-up, product costs, special offer possible markdown or unsold stocks at the end of the season. The FPP may be adjusted by product categories (consistently with the targets described in the strategic plan) and then submitted to the review of the buyers, who, since they cannot deal with high level target such as I and II margins, are able to work on a unit plan level by targeting number of references, purchasing unit costs, sell-out price, discounts, promotions on different types of suppling (planned or *ready to go*). The features of the project allow the analysis of products only up to an article-point-price dimension level because the width and depth of the stock keeping unit level targets are fixed with flat rules linked to the technical box (prepack) defined with vendors.

2.3 *Purchase Plan*

Since buyers are able to define the point-price level for each article, either from a historical or planned basis, in such a framework the FPP turns out to be the starting point of the Purchase Plan. In this area there are utility functions in order to substitute articles and swap of values from one product to another. Throughout historical or planned data analysis, payment delays are defined for each supply source, giving rise to a cash flow plan.

2.4 Financial Check

The plan suggested by the buyers must be approved by the company's management responsible for pursuing financial funds. The latter may send back the plan to the buyers for adjustments or plan according to the FPP. Any further modification of the FPP made by the buyers has an effect on sales, purchases, either by monetary or quantity measures, up to monthly channel-store level, consistently with the strategic plan. Once the FPP has been approved by the finance office, it becomes the final version, giving rise to ERP volume\value per month constraints on purchases by category and product dimensions.

2.5 Buying

During the purchasing phase, buyers may adjust the FPP in terms of number of references, supply unit costs, quantities, articles, price points, delivery months and adapting to current market trends, constantly complying with the fixed margin constraints already fixed at category level. At any time, a buyer may change his own plan being aware of the implications this may have on the *What if* dashboard: indeed, the latter takes into account the approved FPP, the portfolio of purchasing orders currently confirmed and the remaining active balance of the FPP to be purchased. Approaching the selling season of a store, it would still be possible to work on the balance of the FPP not yet purchased, typically involved with *fast ready (always ready)* supplying.

2.6 Financial Forecast

The processes described above generate an economic monthly seasonal forecast for each account closing. Therefore, the finance department is able to simulate variations of the cash flow and levering on credit letter parameters.

3 Results

Process design involves better coordination among different company sectors, consistently with the strategic plan of the firm, without affecting management systems and, guaranteeing a decisional workflow preserving the intermediate targets up to the purchasing phase which maintains some degrees of freedom relatively to the choice of the product, consistently with the objectives defined.

Use of this model has improved performance KPIs like Sell Through and End of season stock, allowing a better margin impact. In addition, it has drastically reduced the planning time for the whole process about 50%, improving handled data quality.

The process was stressed initiating several times the phase margin and sellout plan in the same season sales, being able to better follow the changing market dynamics.

Each new scenario in the margin and sellout model, or in the purchase plan, together with the store development plan, it is performed daily by the Integrated business planning model that provides a new version of the operating income forecast, a balance sheet and a review of cash flow, necessary for proper financial planning.

As a result of the project, the client expects to see an impact on margin of at least 2–5%, from reduction in inventory carrying costs and increased margin on existing sales volume, to new margin on increased sales volume.

4 Conclusion

After the implementation of the Financial Merchandise Planning process each area has an operational need to support day by day, according to any single details coming from operativity.

Managers can engage in decision-making, without bothering to prepare data.

Any impact from the world of sales, purchasing, development of store and product, immediately provides a review of the economic forecast and cash flow, shared with all authorized business functions.

Ongoing activities are aimed at training of users and internal managers to get used to the models of simulation and planning, with the aim not to depend on external consultants in maintenance and parameterization.

The daily use of these models facilitates the work of internal knowledge workers, intended to criticize and continuously improve processes and performance, create knowledge, ways of relating, styles of behavior, being able to shape the organized context of the company.

From this work, several advices were born to implement the model.

In the next step we aim to improve MP models, extending their action scope and considering every requirement of operation functions a company may face in assortment planning, store allocation and sales forecasting.

4.1 Assortment Planning

Both traditional and multichannel retailers struggle with the balance of assortment, breadth and depth of merchandise. Attempting to set the variety of a category, the buyer should consider parameters like profitability of a mix, the corporate philosophy, the physical characteristics of the store, the service level or product availability and the complementary merchandise. The increase of breadth and depth may have a strong (negative) impact on gross margin, requiring a wider range of merchandise on sale: indeed, the more SKUs offered, the greater the chance of breaking sizes. If a popular *sku*, within a given merchandise category, faces a stockout and the buyer cannot purchase it during the season, he/she would typically discount the whole merchandise class. The latter considerations support the importance of linking the FPP with adequate assortment planning, according to store classification (in terms of physical characteristics, geography and buying behavior of their customers), planned backup stock and composition of product vendor prepack (box) in order to estimate the effective GMROI of the FPP and to monitor the buying work in progress.

4.2 Store Allocation and Replenishment

The process of setting and supporting future performance goals for sales, inventory and other financial metrics together with tracking actual results (and variances) regarding those goals may be defined as a replenishment approach. Planning decisions are indeed based on historical trends and management insights into expected future changes, such as number of stores, calendar shifts, business shifts, and promotional events. This approach to merchandise planning, driving company decision making, provides a basis of success measurement with both internal and external parties (ex. stock analysts). However, this also requires user intervention: systems automate the planning process by spreading changes made at higher hierarchy levels down to lower product, location or time levels, or aggregating changes at lower levels back up to hierarchies.

4.3 Promotion and Markdown Management

Product categories typically follow a predictable sales pattern implying introduction, growth, maturity and decline. The shape of the life cycle may be affected by activities of retailers and vendors. It is thus logical to highlight the importance of planning new campaigns, events, advertising and promotions across all channels and analyzing promotional performance in order to improve future strategies. Similarly, this enables planners to set multiple price and markdown structures with complete visibility into inventory, revenue and margin impacts.

4.4 *Social Media Analytics*

Social media provide retailers with an amazing visibility within their customer base: since they are a virtual venue in which agents are able to directly communicate with their customers, they are a powerful tool for collecting and using customers' insights in order to improve planning decisions, through the analysis of social media posting and feedbacks.

References

Bianchini A, Pellegrini M, Peta D, Saccani C (2014) Economic evaluation of investments for workplace safety. Chem Eng Trans 36:49–54. doi:10.3303/CET1436009

Bianchini A, Donini F, Pellegrini M, Saccani C, Fanelli M (2015) Effective implementation measurability in a health and safety management system. In Proceedings of the 25th European Safety and Reliability Conference, ESREL, Zurich, Swaziland, Code 139809, pp 3191–3199. ISBN: 978-113802879-1

Bini V (2011) La Supply chain della moda. Franco Angeli, Milano

BRP (2015) Merchandise Planning Survey

D'Avolio E, Bandinelli R, Pero M, Rinaldi R (2015) Exploring replenishment in the luxury fashion Italian firms: evidence from case studies. Int J Retail Distrib Manage 43(10):967–987. doi:10.1108/IJRDM-07-2014-0098

Granger M, Sterling T (2012) Fashion entrepreneurship. Fairchild Books, New York

Iannone R, Ingenito A, Martino G, Miranda S, Pepe C, Riemma S (2013) Merchandise and replenishment planning optimisation for fashion retail. Int J Eng Bus Manage. doi:10.5772/56836

Kunz GI (2005) Merchandising: theory, principles and practice. Fairchild, New York, NY

la Fleur R (2010) Merchandise planning workbook. Fairchild Books, New York

Levy M, Weitz B (2013) Retailing management. McGraw-Hill

Sacchi S (2015) Fashion puzzle. Franco Angeli, Milano

Yu U-J, Kunz GI (2010) Financial productivity issues related to assortment diversity and supply chain merchandise replenishment strategies. J Fash Market Manage Int J 14(3):486–500. doi:10.1108/13612021011061906

A Proposal for Supply-Chain Improvements in a Luxury Company

Nathalia Tupinambá Karmaluk Tinoco, Ricardo Augusto Cassel and Juliano Denicol

Abstract The fashion market is becoming more competitive in the luxury industry, mainly because of the appearance and strengthening of other industries that provide new products more consistently to its customers. This sector, marked by its two annual collections and its unique items, must become more flexible and provide a faster response to its consumer base. In this context, supply chain management is an important factor in determining the degree of competitiveness of a company. This paper proposes tools to assist in the production scheduling of a luxury industry through making its supply chain more flexible by reducing the number of items delivered before the release, thus allowing the company to react according to the sales. To achieve that, an approach to the calculation of orders based on sales forecasts and on the lead time between stores and distribution centers was developed, in addition to a tool to control the delivery of the items through better integration with the suppliers. The proposed approach reduced the number of items in the first two orders by 17 and 12%, and several delays were avoided through the utilization of the order-control tool.

Keywords Supply chain management (SCM) · Integration with suppliers · Distribution requirements planning (DRP) · Fashion accessories · Luxury company

1 Introduction

Trends in the fashion market have changed in recent years. The luxury business sector, which before relied on customer's loyalty to their products, saw these same clients be seduced by the fast-fashion sector. The capacity to be able to constantly

---

N.T.K. Tinoco (✉) · R.A. Cassel · J. Denicol
Industrial Engineering Department, Federal University of Rio Grande do Sul (UFRGS), Porto Alegre, Brazil
e-mail: nathalia.karmaluk@gmail.com

R. Rinaldi and R. Bandinelli (eds.), *Business Models and ICT Technologies for the Fashion Supply Chain*, Lecture Notes in Electrical Engineering 413,
DOI 10.1007/978-3-319-48511-9_19

offer new products to the market, such as is the case of the Spanish brand Zara, increased competitiveness in the fashion market, forcing luxury companies to seek ways to adapt to this change (Reinach 2005).

Luxury firms are characterized by the importance given to the brand name and its exclusivity as a way of attracting customers. Fast fashion, on the other hand, is characterized by quick responses to the consumer and to the changes in fashion trends. Although these two branches of the fashion market have almost opposite characteristics, luxury companies have been changing in order to become more agile, like the fast-fashion companies. Among these changes is the release of a pre-collection with some exclusive pieces, which is an addition to the two annual collections (Reinach 2005).

In this context, an essential topic is supply chain management, which, in the current competitive world, has become one of the most important factors in determining the degree of competitiveness of a company (Slack et al. 2007). It is vital to understand that, in order to have an efficient supply chain, it is not enough to have the latest technology and optimized processes, but that it also requires skilled and communicative professionals. Giunipero et al. (2006) stated that these professionals should focus not only on reducing costs, but also on their company's integration and collaboration with suppliers.

The objective of this study is to offer tools to aid in the supply process of a fashion-sector company in order to make its supply chain more flexible. When dealing with suppliers, one of the key points is the order placement by the buyer (Giunipero et al. 2006). This paper presents a case study of a company where the calculation of the orders sent to suppliers was modified to include the lead time between the different storage centers in order to achieve a more approximate calculation of the required quantities of products. Thus, current flexibility is expected to increase, which will allow the company to react according to the sales, ordering more or less of each product according to the observed demand.

Another point to be considered is that, by including the lead time between storage centers, the safety margin that existed before is lost. As a result, suppliers will not be able to permit delays to occur, because the calculation has become more accurate. Because the communication between the production team and suppliers becomes essential, with the development of informatics and telecommunication networks, there is a need for continuous control of information in order to process and transmit it quickly, in order to gain a competitive advantage and maintain a good level of efficiency.

2 Literature Review

Slack et al. (2007) addressed the difficulty of separating production planning from control activities. Both try to conciliate the market demands and what operations can provide, however there are some characteristics that distinguish them. Planning, for instance, addresses what is expected to happen in the future without the guarantee that this event will happen, and therefore it is exposed to variations. The control, on the other hand, is used to make the necessary adjustments in order to mitigate the effects of these variations, based on indicators that detect what is happening in the system.

Due to the large competition between companies, supply chain management has become essential for an organizations' survival (Slack et al. 2007). According to Rollins et al. (2003), it is possible to create a general model of a supply chain that can be used as a basis for a more specific model for companies in the same sector. This reference model contains the elements that are common to the sector's operations. For the clothing and accessories sector, for instance, the authors stated that the chain begins with product design, going through production planning, manufacturing, and, ultimately, distribution of products to the stores.

Kim (2009) indicated that supply-chain integration plays a significant role regarding the supply chain management strategies, the capacity for competition, and the company's performance. In the early stages of the integration process, the level of integration has a direct role in the company's performance. However, as this process is implemented, it can be noticed that the level of integration has a direct role in the relationship between supply chain management and the capacity for competition, and an indirect one on the company's performance.

In that way, even if a company has great supply-chain-management practices and competitive ability, it is essential for its performance to have integration with suppliers. This is confirmed by Vickery et al. (2003), who added that information technology also has a direct role in the integration of the chain, which influences the customer service, and is ultimately linked to a company's performance.

Flynn et al. (2010) analyzed supply-chain integration through three dimensions: (i) internal, between the sectors of the company; (ii) with consumers; and (iii) with suppliers, concluding that the internal integration forms a basis for integration with customers and suppliers. Thus, it is suggested that a company starts the integration process internally and afterwards moves on to the other two dimensions. Furthermore, Zhou et al. (2014) conclude that, to achieve good performance, it is necessary to have quality in the information exchanged among participants, which enables cost reduction for data collection and production control and increases the confidence level between the parties involved, a key factor for integration.

In the scope of the integration of a company with its suppliers, there are two flows, in opposite directions. The first is the flow of materials from suppliers to the company, considered as a "*forward*" integration. The second is the information flow from the company to suppliers, which is considered a "*backward*" integration. The importance of these two flows provides both opportunities and challenges for the companies, mainly due to the current trend to outsource activities that are not directly related to the core competence of the company. This trend emphasizes the need for greater integration, through the use of precise and current information (Prajogo and Olhager 2012).

Van der Vaart and Van Donk (2008) propose three categories through which it is possible to analyze supply chain management: practices, standards, and attitudes. Supply-chain practices can be defined as the activities or technologies that play an important role in the collaboration between the company and its suppliers or consumers. The standards of the interaction can be determined through the frequency in which information is exchanged. The attitudes, on the other hand, are related to the expectations about the relationship between the two parts. The authors conclude that, although the literature shows that there is a direct relationship between integration and company's performance, the study of these three categories enables better understanding of the relationship between suppliers and buyers. Prajogo and Olhager (2012) add that long-term relationships between companies and suppliers positively influence the exchange of information and communication between the two parties, as well as the company's performance. Also, the frequency, quantity, and quality of information are the main factors of this integration.

According to Bruce et al. (2004), for clothing and accessories companies, integration and good relationships with suppliers are the main factors to ensure their supply-chain flexibility. This is because these companies must be able to respond quickly to market changes, but cannot carry large inventory, because their products have a short life cycle, usually six months to a year.

3 Methodology

This study was carried out in five steps: (i) data collection; (ii) parameterization of the products information in order to use a Distribution Requirements planning (DRP) program; (iii) creation of a tool to perform the orders' calculation using the new approach proposed; (iv) comparison of the results obtained in both approaches; and (v) creation of a tool to perform the orders delivery control.

4 Results

In this section, the results are presented after the application of the proposed methodology in this scheme: (i) the creation of different tools, using a new approach; and (ii) the validation of these tools, through comparison with previously used approaches.

4.1 Data Collection

Data were collected regarding the products of the Fall-Winter 2013 collection, such as the registration code, release date, category, and supplier of each material, WDs safety stocks, which regions would receive a particular product, and the lead time between these regions and the WDs. These data were used to calculate the weekly demand at the stores, RWDs and CWD.

In addition, orders made for the collection were gathered, and were calculated with the former approach, where the different lead times between the stores and the WDs were not considered. Thus, many items were delivered ahead of time to regions nearby the CWD, or delayed for the most distant regions of the CWD.

4.2 DRP Program Utilization

The company's DRP program calculates how many parts must be ordered from the supplier per week, in order to not cause stock disruptions at the stores. This calculation is based on the sales forecast, on the amount of parts in stock at the CWD, and on lead times between the distribution centers. Although this tool cannot be used during the collection for seasonal products, since orders are made months before the release, it can be used to react according to the sales.

When there are no deliveries, and therefore, no stock at the CWD, the DRP calculation are based on the sales forecast and the minimum items necessary in every store, every week. Then, according to the lead time between the store and the RWD, this number of items needed is transferred—"x" weeks to RWD. Finally, the same happens between RWD and the CWD with the lead time of two weeks. In Fig. 1, for instance, the first lead time is equal to five weeks, and therefore, if they need to have 100 parts in the store in a given week, the items need to be at the RWD five weeks prior and at the CWD seven weeks prior.

A faulty point in this program is, however, that it has the correct amount of lead time only for the distribution between the CWD and the RWDs, considering it to be zero between RWDs and the stores. Thus, a tool was created in Excel that adjusts the calculation using the true value, in order to calculate the correct demand at the CWD. Since the company has been in business for a long time and controls the entire supply chain, lead times are known and suffer little or no variation, being two weeks between the CWD and the RWDs and ranging from one to eight weeks for the RWD to the stores.

In the tool developed, based on the product's information, Excel extracted information relating to the total quantity to be requested from the supplier, to the CWD and RWDs safety stocks, and to the lead time between WDs and the stores.

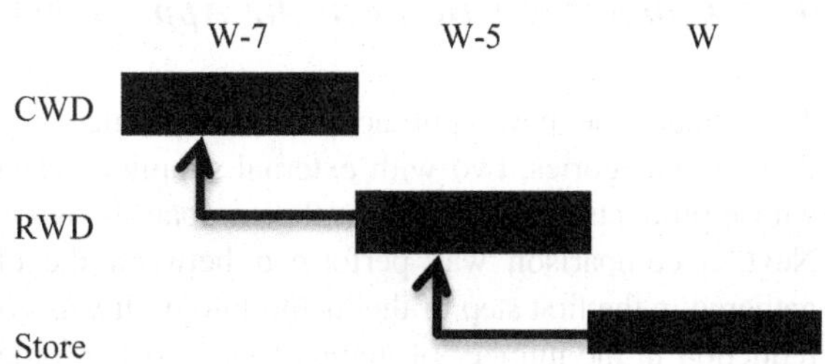

Fig. 1 Lead time influence on demand between WDs and the stores

Next, the weekly demand per store was copied from the DRP and linked with the right lead time, ending the existing gap and providing the correct calculation.

In addition, the DRP calculates that all the WDs safety stock must be delivered in a week, but the company's policy is to separate this amount in a period as long as four weeks, mainly to reduce losses from theft or damage to the shipments. Thus, the number of parts referring to the safety stock was copied from the program, representing the amount that should be at the WDs in each week. To that information, a certain percentage is applied by the company in order to distribute it over more weeks.

These two results, the correct demand in stores and the safety stock divided into shipments covering more than one week, were then copied to the DRP. Based on the sales forecast and lead times, the result was found for the quantity of pieces that should be at the distribution centers each week in order to avoid delays in deliveries to stores.

4.3 Orders Calculation

Based on the real need for items at the CWD, it was possible to calculate the orders that were to be made to suppliers, which then established the delivery dates. These were determined in relation to the release date of each product, but usually five deliveries are made: (i) six weeks before the release; (ii) two weeks before; (iii) two weeks later; (iv) six weeks later; and (v) eight weeks later. However, since it was observed that some managers vary the scheduled weeks, a tool that allows them to vary for each product was developed.

With the delivery dates defined, the orders were generated through the aggregation of the CWD weekly demands between deliveries. For example, if the first order must be delivered at week 30 and the second at week 34, the first must contain a sufficient amount of items so that the CWD does not have a disruption before the arrival of the second order, until the end week 33. All orders were calculated following this logic, except the last. As each product has a maximum number of items produced per collection, the last order was the difference between this amount and the sum of the other orders.

4.4 Comparison Between the Approaches

To validate the new approach, the order calculation tool was applied to three different categories, two with external suppliers, categories "X" and "Z", and one whose products are produced at the company's own workshops, the category "Y". Next, a comparison was performed between the obtained results and the data gathered in the first step of the methodology. It was possible to note that there was a reduction in the number of items of the first two orders and an increase in the last three, being higher in the last one. Figure 2 shows the number of pieces per order of each approach.

Tables 1 and 2 show, respectively, the comparisons per unit and the percentage between the two approaches, as well as the delivery reduction for each category. In general, there was an improvement with the tool use, since there was a decrease in the first orders. This was relevant because, prior to the first order receipt, twenty units should have undergone a quality test, and the order could not be delivered until these twenty pieces were approved. For instance, if three of the twenty were not accepted, then three pieces must be resent until a total of twenty have passed the test. This fact generates many delays, because it can take a while until they are all accepted and, if changes in manufacturing are necessary, all of the first order pieces that are ready up to that moment must go through rework. In addition, the increase in the number of items delivered after the launch was one of the program's goals, as it makes the supply chain more flexible, enabling changes to the amount based on the first weeks sales.

In Table 2, when the categories are separate, it is observed that the behavior of the category "Y" is not equal to the others. This difference is due to the fact that the

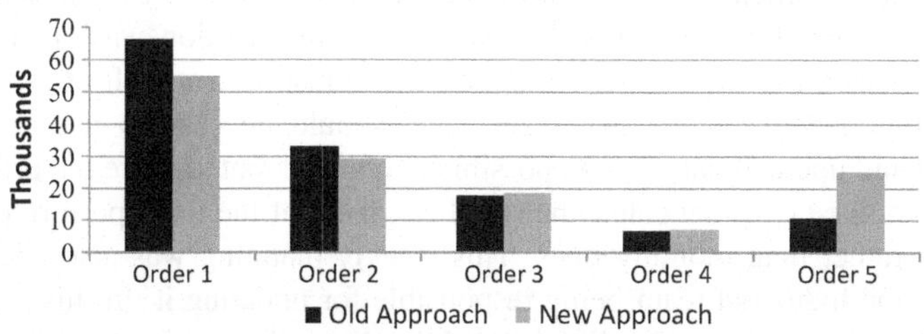

Fig. 2 Comparison between the old and the new approach

Table 1 Comparison between the two approaches

Approach	1	2	3	4	5	Total
Old	66325	33166	17765	6796	10609	134661
New	54827	29213	18462	7056	25103	134661
	−17%	−12%	4%	4%	137%	

Table 2 Reduction per category

Category	Supplier	1 (%)	2 (%)	3 (%)	4 (%)	5 (%)
X	Extern	−49	−13	−39	−10	592
Y	Intern	12	12	−15	−35	−11
Z	Extern	−18	−19	66	108	539
General		−17	−12	4	4	137

relationship between the production team, responsible for making the orders, and the industrial team, representatives of the suppliers, is different between such categories. For the "X" and "Z", since they are external to the company, orders were made monthly and rarely changed. In the "Y" category, as the industrial team is the provider, the orders, even in the old approach, were reviewed weekly and modified in accordance with the deliveries or sales, if the product had already been launched, which enabled a smaller amount to be requested at the beginning, since it could be modified later.

Similarly, it was noticeable that certain products, within the categories "X" and "Z", experienced an increased in items in the first order. Since they are exact cases, it was concluded that, with the previous approach, a demand lower than the real one was calculated.

4.5 Orders Delivery Control

After the orders are calculated and passed on to suppliers, the control of the deliveries is fundamental by comparing them both with the pre-established date and with the demand. For such, a tool to gather this information was created, which must be updated by the production and by the industrial teams. It aims to achieve the integration between the two teams, and should be easy to use and contain important and updated data. It was possible to observe that, despite the orders being placed with fixed delivery dates, the suppliers, most of the time, performed weekly deliveries rather than monthly ones. This weekly planning was also added to the tool, with the industrial team being responsible for updating it. In this way, it was possible to compare the real deliveries and the demand weekly, and the cumulative deliveries and the requested orders monthly.

Figure 3 shows the overall situation concerning the collection the accumulated data. The gray and black vertical lines represent, respectively, the week in which the file is being updated and the release date. The dashed gray line shows the orders that were requested. The thick black line shows the parts that have already been delivered, and the dashed black line shows the vendor's planning for future deliveries. The area in dark gray indicates the demand, and the area in light gray, the safety zone. The latter is equal to the dark gray, shifted two weeks, and its purpose is to draw the attention of the production and industrial teams for when the black or gray lines, of orders and deliveries, enter this area.

After being developed, the tool was tested with three industrial staff members as a pilot, with the purpose of finding points for improvements. There was no training, and all explanations were conducted via video conferencing, which proved the easiest to use. An example of the tool use can be seen in Fig. 3. It is easy to see that the order of the second week of 2014 (S201402) could not have delays because it would result in a delay in the items' delivery to the stores, since there is no margin between the order quantity and the demand. Observing this, the supplier changed its planning to advance the delivery of this order, providing a higher margin.

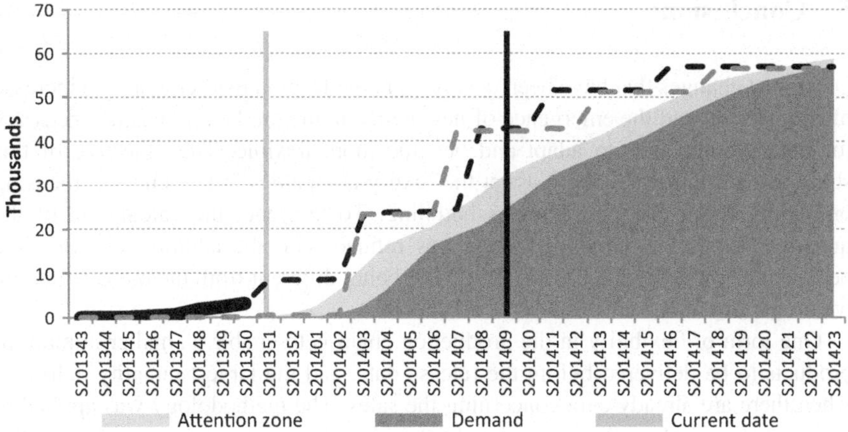

Fig. 3 Collection situation

One of the improvements requested by members, who have utilized the tool, is the service-level calculation, which is one of the indicators most commonly used by the company. The required data were already on the chart, since it is given by the relationship between full black lines and the gray dashed line, i.e., deliveries performed and orders. Therefore, a field highlighting this information was added. Moreover, as this tool is updated weekly, suppliers may change the deliveries planning and improve this percentage from one week to another.

Another type of information added after the tests were performed was the items backlog. With the schedule of deliveries information from suppliers and the quantity of parts delivered, it was possible to calculate how many items were missing and how many extra items were delivered. With that, suppliers can increase or decrease the number of items to be delivered in the following week in order to not damage the service level.

Since the tests with the three members of the production team occurred without major problems and they quickly adapted to the tool and to its routine of weekly updates, the tool stepped out of the pilot mode and was applied to all categories. A meeting with all members of the two teams was organized where all the points were explained, from the change in the orders calculation up to the pilot project conduction, emphasizing that collaboration between the two teams was essential.

The implementation to all products occurred rapidly, and everyone adapted to the routine of the tool, perceiving its importance. By talking to team members, it was observed that they understood the objective of maintaining better control of supplier deliveries, and that the tool optimized this control and avoided many delays that would probably not have been avoided with the previously used tools. The old tools were updated less frequently, and, therefore, it sometimes took too long to notice a delay, with the situations already being irreversible by the time they were discovered.

5 Conclusion

Due to the changes that have been occurring in the fashion market, with the increase of competition and the emergence of new items on the market, it became necessary for luxury companies to adapt and become more flexible. The objective of this study was to propose tools to assist the luxury companies in this change, focusing on supply chain and production scheduling. To this end, the calculation of the quantity of parts required per week was refined with the addition of lead time between the different RWDs and the CWD, being different from the old calculation, which assumed the same value for all RWDs.

One form of flexibility in the production chain is to enable a smaller number of products to be delivered before the release date, and the largest quantity delivered when there are already data concerning the sales. The methodology was applied in three categories of products, for suppliers with different characteristics. In the "X" and "Z" categories, which are external suppliers to the company, there was a reduction of 49% and 18%, respectively, in the first order amount of items. In the "Y" category, which was an internal supplier, there was a 12% increase in this order. This is due to the fact that, in the latter, the number of items needed was reviewed weekly, and modified in accordance with the deliveries and sales, while in the first two, the orders were made months before delivery and were rarely changed.

With a more refined calculation of the amount of items, there is less margin for error, and therefore better and stricter delivery control of orders is necessary. Thus, a tool was developed to take into account the number of parts required per week, the orders dates, amounts requested, and the delivered orders. The reaction obtained from the managers of the three categories was positive, and so the tool was easily implemented and became a part of their routine. In addition, it was observed that the tool promoted greater integration between the teams, avoiding possible delays which were identified early enough to be prevented.

Future developments include a comparison between the percentage of delay per collection, both before and after the implementation of the order control tool. Moreover, the estimation of cost reduction through lower inventories and reduced delays constitutes a topic of interest, as well as does the consideration of stochastic lead times between WDs.

References

Bruce M, Daly L, Towers N (2004) Lean or agile: A solution for supply chain management in the textiles and clothing industry? Int J Oper Prod Manage 24(2):151–170

Flynn BB, Huo B, Zhao X (2010) The impact of supply chain integration on performance: A contingency and configuration approach. J Oper Manage 28(1):58–71

Giunipero LC, Handfield RB, Eltantawy R (2006) Supply management's evolution: key skill sets for the supply manager of the future. Int J Oper Prod Manage 26(7):822–844

Kim AW (2009) An investigation on the direct and indirect effect of supply chain integration on firm performance. Int J Prod Econ 119(2):328–346

Prajogo D, Olhager J (2012) Supply chain integration and performance: the effects of long-term relationships, information technology and sharing, and logistics integration. Int J Prod Econ 135 (1):514–522

Reinach SS (2005) China and Italy: Fast fashion versus prêt a porter—towards a new culture of Fashion. Fash Theory 9(1):43–56

Rollins RP, Porter K, Little D (2003) Modeling the changing apparel supply chain. Int J Cloth Sci Technol 15(2):140–156

Slack N, Chambers S, Johnston R (2007) Operations management. Prentice Hall, Harlow

Van Der Vaart T, Van Donk DP (2008) A critical review of survey-based research in supply chain integration. Int J Prod Econ 111 (1):42–55

Vickery S, Jayaram J, Droge C, Calantone R (2003) The effects of an integrative supply chain strategy on customer service and financial performance: an analysis of direct versus indirect relationships. J Oper Manage 21(5):523–539

Zhou H, Shou Y, Zhai X, Li L, Wood C, Wu Z (2014) Supply chain practice and information quality: a supply chain strategy study. Int J Prod Econ 147:624–633

Part IV
Sustainable Fashion Supply Chain

Integrating Sustainability in the Fashion System Using Association Rules

Filippo Emanuele Ciarapica, Ilaria De Sanctis, Barbara Resta, Stefano Dotti, Paolo Gaiardelli, Romeo Bandinelli, Virginia Fani and Rinaldo Rinaldi

Abstract This work aims at analysing the environmental strategies developed by fashion companies in order to identify the most important stakeholders involved in the strategy, the environmental practices carried out and the competitiveness impacts. A conceptual model, based on Association Rules (AR), has been proposed for investigating the network of influences among the environmental strategy, the environmental management practices and the environmental competitiveness and profitability of the companies. The research has been conducted through a survey submitted to fashion companies operating in Textile, Clothing and Leather (TCL) sectors. Results indicate that "customers" and "suppliers" are the most

F.E. Ciarapica · I. De Sanctis
Dipartimento di Ingegneria Industriale e Scienze Matematiche,
Università Politecnica delle Marche, Via Brecce Bianche, 60100 Ancona, Italy
e-mail: f.ciarapica@univpm.it

I. De Sanctis
e-mail: i.desanctis@univpm.it

B. Resta · S. Dotti · P. Gaiardelli
CELS—Research Group on Industrial Engineering, Logistics and Service Operations,
Department of Management, Information and Production Engineering,
University of Bergamo, Viale Marconi 5, 24044 Dalmine (BG), Italy
e-mail: barbara.resta@unibg.it

S. Dotti
e-mail: stefano.dotti@unibg.it

P. Gaiardelli
e-mail: paolo.gaiardelli@unibg.it

R. Bandinelli (✉) · V. Fani · R. Rinaldi
Department of Industrial Engineering, University of Florence,
Viale Morgagni 40, 50134 Florence, Italy
e-mail: romeo.bandinelli@unifi.it

V. Fani
e-mail: virginia.fani@unifi.it

R. Rinaldi
e-mail: rinaldo.rinaldi@unifi.it

R. Rinaldi and R. Bandinelli (eds.), *Business Models and ICT Technologies for the Fashion Supply Chain*, Lecture Notes in Electrical Engineering 413,
DOI 10.1007/978-3-319-48511-9_20

important stakeholders to be involved in the environmental strategy. A greater compliance and a risk reduction can be reached when the Financial institutions, Environmental NGOs, Rivals and Shareholders are involved in the environmental strategy. In this regards, results highlighted the importance of Environmental auditing programme for suppliers and Sustainable disclosures. Moreover, in fashion sector, the companies that pursued "Lower cost" as competitive advantage aim at obtaining from environmental strategy a great access to capital and lower cost of capital.

Keywords Environmental sustainability • Textile, clothing and leather (TCL)

1 Introduction

The Textile, Leather and Clothing (TCL) sector represents one of the most important economies in the European Industry and it contributes to a large extent to the global fashion system: consistent investments, emphasis on upmarket products and global leadership have all made Europe the world's biggest exporter of treated leather, as well as the second biggest exporter of textiles and clothing, with respectively 24% and 26% of world sales (Euratex 2014). In 2014 the textile and clothing industry realised a turnover of €165.3 Billion and employed 1.6 Million people in more than 170,000 companies (Euratex 2015). The leather industry represented in 2011 a turnover of €31 Billion and consisted of about 24.000 SMEs and employs 392.264 people (ESSC TCLF 2014). Within this context, Italy is the principal actor of this economy with 18.7 billion Euros of value added and more than a third of the total EU production (Eurostat 2013a, b, c).

At the same time the TCL industry is recognised as one of the most polluting sector in the world, being a huge consumer of water, electricity and chemicals, and discharging massive quantities of wastes to land (Bevilacqua et al. 2014). Therefore, the TCL industry is one of the sector that has received the highest public attention, and one of the most challenged by the sustainability concerns that have recently emerged in the media, including Detox by Greenpeace, Sustainable Apparel Coalition, Clean Clothes Campaign, Røadmap to Zero Discharge Of Hazardous Chemicals (ZDHC), Fashion Revolution, only to mention a few.

In this context sustainability initiatives are crucial for fashion firms to increase customers' and stakeholders' engagement and loyalty and stay competitive in the market (Smith 2003). However, there is still a dichotomy between today's reality and the industry's opportunities to leverage sustainability, as well as between the rhetoric and the reality in pursuing a sustainability approach (Rhee and Lee 2003): while many companies commit to sustainability, few put their commitment into actions (Chi 2011). In literature some preliminary results have been presented on how and to what extent TLC companies implement sustainability practices along their supply chain (De Brito et al. 2008; Chi 2011; Caniato et al. 2012; Resta et al.

2014), but a comprehensive analysis encompassing the all the corporate environmental management areas is still missing.

Thus, the aim of this paper is to contribute to enrich this research field, analysing how the implementation of environmental strategies and practices boost corporate competitiveness in the Textile, Clothing and Leather (TCL) sector.

The paper is structured as follow. Section 2 presents an overview of the theoretical background, while in Sect. 3 the research methodology is explained. Section 4 reports the results of this work. Finally, in Sects. 5 and 6 discussion of results, conclusions and the future work are described.

2 Theoretical Background

Over the past decades, the issue of environmental sustainability has steadily gained greater prominence on the corporate agenda of TCL companies, resulting in an increasing research interest in analysing the implementation of a corporate environmental management framework. In particular, corporate environmental management encompasses all the strategic and operational efforts to minimize the negative environmental impact of firm's business operations (Cramer 1998). In line with "traditional" corporate management, environmental management become manifest through an integrated approach encompassing (Carmona-Moreno et al. 2004): (i) "environmental strategy", that denotes a firm's competitive orientation towards the environmental pillar of sustainability; (ii) "environmental management practices" that refer to all the strategic, tactical and operational activities which aim to protect the environment; and (iii) "environmental competitiveness and profitability" that denotes the contribution of a company's sustainability management to its overall competitiveness.

Environmental strategy. A key point in the discussion of environmental strategies is the fit between environmental strategy and corporate competitive strategy (Baumgartner and Ebner 2010), as well as their integration (Galbreath 2009; Murthy 2012): any sustainability approach that is "fragmented" and "disconnected from business and strategy" will "obscure many of the greatest opportunities for companies to benefit society" (Kramer and Porter 2006) and to increase company's alignment with the expectation of its stakeholders (Henriques and Sadorsky 1999).

Moreover, different types of environmental strategy could be implemented by a company, based on their proactivity level that typically ranges between two extreme positions: environmental reactivity, representative of companies that only implement the minimal compulsory changes to meet regulations, and environmental proactivity, typical of companies that voluntarily take measures to reduce their impact on the natural environment. Several works conceptualised this path from reactivity to proactivity by establishing a number of progressive stages, assuming a single and linear path that companies follow when developing their commitment to the natural environment (Jabbour 2010).

Environmental Management Practices (EMPs) refer to all the measures and activities aimed at executing an environmental strategy, thereby reducing the environmental impact caused by a company's business (Sroufe et al. 2002). The practices mostly applied in the TCL industry can be classified into five categories (Resta et al. 2014): product, supply chain management, process, governance and culture.

The first category includes all the practices related to the design of sustainable products, as well as the required raw materials. Practices which can be included in this dimension are the systematic adoption of environmental assessment tools (e.g. LCA, Carbon Footprint, Water Footprint) related to the final product, the utilisation of methodologies to improve the product design and development process (e.g. Eco Design), as well as products certifications (i.e., Oeko-Tex, Seri.Co, Ecolabel, GOTS, etc.) and in general practices for the improvement of product environmental sustainability. Consequently, the "Product" variable includes also actions related to the use of raw materials recycled from waste, produced from renewable sources, close to the sourcing point (e.g. zero km raw materials suppliers), and/or certified. Practices related to the reuse of packaging are also considered.

The second category deals with practices related to the establishment, management and control of a green supply chain, whereby the key dimensions of corporate sustainability are combined with supply chain characteristics. Evaluation of environmental performance of suppliers, verified by environmental auditing programme, has received the highest attention in supplier selection for sustainable supply chain management. Another important practice in this area refers to the implementation of collaborative relationships with suppliers in order to improve the environmental sustainability levels of purchased materials and technologies. With reference to transportation, the use of hybrid or electric transport systems is also included.

The third dimension is related to production processes. In this area, practices related to energy, water, process materials and waste are included. Moreover, process and environmental management system certification (e.g. ISO 14000, EMAS) are considered as relevant.

The fourth dimension refers to initiatives aiming at managing green activities and the relationship with all the stakeholders. Sustainability reports, Sustainability Advisory Board, dedicated corporate function or business unit, website dedicated to green activities are practices considered in this dimension.

Finally, the culture dimension deals the with introduction of training programs involving both internal than external stakeholders, with the goal of spreading a green culture.

Environmental competitiveness and profitability. Savitz and Weber (2006) suggest that "a sustainable company is one that creates profit for its shareholders while protecting the environment and improving the lives of those with whom it interacts". Indeed if environmental issues are successfully incorporated into corporate strategies and practices, sustainability can be transformed into competitive advantage, making a business case for environmental sustainability (Epstein and Roy 2003; Salzmann et al. 2005).

Empirical research has very much focused on the relationship between environmental strategy and business performance, offering contradictory results. As reviewed by Blomgren (2011), while some studies demonstrated that firm's environmental strategies lead to improved business performance, others found either insignificant or negative relationships. Moreover, considering the fashion sector, while there is growing evidence of the use of environmental management practices by TCL companies (Caniato et al. 2012; Choi et al. 2014; Cooperrider and Fry 2012; Na and Na 2015; Resta et al. 2014), there is still a selective evidence on the benefits that might be obtained. Lo et al. (2012) focuses on the adoption of a specific practice only (adoption of ISO 14000 certification) and its impact on firms' financial performance, confirming a significant positive impact on ROA and ROS. Therefore, the argument of whether or not environmental strategies lead to improved business performance is far from resolved and further empirical research is thus needed to understand the link between environmental management practices and business competitiveness.

3 Research Methodology

In order to analyse how the implementation of environmental strategies and practices boost corporate competitiveness in the Textile, Clothing and Leather (TCL) sector, an explanatory survey research has been conducted (Malhotra and Grover 1998). In theory building research, no matter how inductive the approach, a prior view of the general constructs or categories and their relationships is valuable in order to: (i) shape the initial research design, (ii) measure constructs more accurately, and (iii) have a firmer empirical grounding for the emergent theory (Forza 2002). Therefore, a set of variables of interest have been identified for each corporate environmental management dimension, as described in the Theoretical Background.

The primary data for this study was collected through a web survey in June–July 2014. The online survey was chosen among different ways to administer the questionnaire to avoid costs, reduce bias and automate data. The research was initiated with the design of the research model and the related questionnaire, which was the main source for data collection. A pilot-test questionnaire with a set of selected companies was administered to prove viability and to detect difficulties in the interpretation of questions. The inputs received from the pre-test were analysed and weighted in the final version of the questionnaire. It comprised 27 questions and was intended to elicit information on environmental strategy, management practices and competitiveness. In particular, the questionnaire was categorised into 4 sections:

- Section A—Company's general information
- Section B—Environmental strategy
- Section C—Environmental management practices (EMPs)

- Section D—Environmental competitiveness

The questionnaire was administered to a subset of companies operating in the TCL Italian sector included in the AIDA data base (NACE code: 13, 14, 15). From the total population of companies having 13, 14 and 15 NACE code, the present study investigates the firms that have explicitly declared on their website the adoption of almost one sustainability practice, totaling 514. Each company was contacted by email, addressed to potential respondent managers knowledgeable about the phenomenon to be measured, and reporting the link to the web-questionnaire. Follow-up telephone calls after 2 weeks resulted in 343 total usable responses returned to the authors, corresponding to a response rate of 67%.

The collected data were then analysed through the association rule analysis, as described in the following paragraph.

3.1 Association Rule Analysis

Agrawal et al. (1993) introduced association rules for discovering regularities between products in large-scale transaction data recorded by point-of-sale systems in supermarkets. Following the original definition of Agrawal et al. (1993) the problem of association rule mining is defined as: let I = $\{i_1, i_2, \ldots, i_n\}$ be a set of n binary attributes called items, and let D = $\{t_1, t_2, \ldots, t_m\}$ be a set of transactions called the database. Each transaction in D has a unique transaction ID and contains a subset of the items in I. A rule is defined as an implication of the form A $\rightarrow$ B, where A and B are known as item sets, and A, B $\subseteq$ I, A $\cap$ B = $\emptyset$. Agrawal et al. (1993) considered each rule composed by two different sets of items, called item-set (A and B). Hence, A is named the antecedent or left-hand-side (LHS) and B is the consequent or right-hand-side (RHS). Moreover in order to select rules of interest, three common metrics (support (1), confidence (2) and lift (3)) are defined for measuring association between the antecedent and the consequent.

$$Support = \frac{\#\{LHS \cup RHS\}}{\#\{all\, records\}} \tag{1}$$

$$Confidence = \frac{Support\,\{LHS \cup RHS\}}{Support\{LHS\}} \tag{2}$$

$$Lift = \frac{Support\,\{LHS \cup RHS\}}{Support\,\{LHS\} \times Support\,\{RHS\}} \tag{3}$$

Therefore, the “Support” indicates how frequent combination of item-sets occurs in the dataset for an alternative. In contrast, the confidence is equivalent to the conditional probability, the probability of finding the RHS of the rule in transactions under the condition that these transactions also contain the LHS. Furthermore, the lift tells us whether the antecedent and the consequent are independent (zero), positively correlated (above unity), or negatively correlated (below unity).

4 Results

4.1 Sample Characteristics

The sample is mostly composed by companies belonging to the textile sector (88%). The remaining part is equally divided into clothing (6%) and leather (6%) companies. Moreover, 7% of the sample companies have turnover higher than 50 million € (large companies), 24% are medium-size companies (50 mil.€ < Turnover < 10 mil.€) and 69% are small and micro-size companies (Turnover < 10 mil.€).

4.2 Association Analysis

Association Rule Mining is performed on the data set (343 cases) and 58 items in the data set using the Apriori algorithm implementation from the arules package for R. In order to determine relevant mined rules, only rules with a support greater than 0.9 have been considered. Indeed, as the support measure evaluates the probability that items occur, probability greater than 90% has been taken into account.

We imposed the different levels of firms' competitive strategies (Product differentiation, Cost Leadership and Hybrid), firms' competitive advantage due to sustainability practices (Revenue growth, Cost Saving and Compliance and risk) and "strategic behaviour", as right-hand-side (RHS) item-set, in order to investigate which factors mainly influence these aspects. All other variables have been inserted as left-hand-side (LHS) variables.

The RHS and LHS variables considered are mainly dummy variable, as they take value 0 or 1 to indicate the absence or presence of the effect. In addition, a five point Likert scale represents the variable "Integration of environmental and competitive strategy" and measures the proactivity level of a company.

Tables 1 and 2 report the results of the analysis ordered in decreasing order by confidence. These rules have satisfied the minimum support (0.9) and minimum confidence (0.9) along with lift value greater than one. The support and confidence columns indicate that there is both a frequent combination of item-sets and a high conditional probability of finding the RHS of the rule in transactions under the condition that these transactions also contain the LHS. Moreover a Lift greater than 1 confirms that the variables are positively correlated. Significant values have been found only with the Compliance and risk (Table 1) and Cost leadership variables (COMPADV_COST) (Table 2).

Table 1 Association rules output (Consequent "Compliance and risk")

Rule no.	LHS	=>	RHS	Support	Confidence	Lift
1	{$STAKE\_INT\_PRIM_3$ = 1, $STAKE\_SECO_2$ = 1}	=>	{Compliance and risk = 1}	0.908	0.989	1.6
2	{$STAKE\_SECO_2$ = 1}	=>	{Compliance and risk = 1}	0.960	0.986	1.3
3	{$STAKE\_INT\_PRIM_3$ = 1}	=>	{Compliance and risk = 1}	0.931	0.986	1.3
4	{$STAKE\_SECO_2$ = 1, $PRACT\_TRANS_2$ = 1}	=>	{Compliance and risk = 1}	0.921	0.986	1.2
5	{$STAKE\_SECO_2$ = 1, $PRACT\_RAWM_3$ = 1}	=>	{Compliance and risk = 1}	0.921	0.986	1.2
6	{$STAKE\_INT\_PRIM_2$ = 1, $STAKE\_SECO_2$ = 1}	=>	{Compliance and risk = 1}	0.911	0.986	1.2
7	{$STAKE\_SECO_1$ = 1}	=>	{Compliance and risk = 1}	0.999	0.986	1.2
8	{$PRACT\_SUPPLY_2$ = 1}	=>	{Compliance and risk = 1}	0.908	0.986	1.2
9	{$PRACT\_GOV_1$ = 1}	=>	{Compliance and risk = 1}	0.901	0.986	1.2

5 Discussion and Implications

Many rules defined in the results section can be used by managers in order to develop the best strategy for integrating environmental and competitive strategy. To sum up and explain the results obtained with "Compliance and risk" as a consequent (Table 1), it is possible to state that:

- Rules 1, 2, 3 and 6 of Table 1 mean that compliance and risk advantage is most likely to be reached when the Financial institutions, Environmental NGOs and Shareholders are involved in the environmental strategy. The study highlights a strong connection between finance and the environment. The environmental risks that confront a financial institution's clients such as violation of laws and responsibility for cleaning up contamination have an impact on their bottom line and, in turn, can pose risks to the institution and shareholders.
- Rules 4 and 5 mean that compliance and risk advantage is most likely to be reached when Environmental NGOs are involved in the environmental strategy, while "Zero km" raw material and Low impact vehicles are the environmental practices carried out. Most Environmental NGOs recognise the benefits of the wide variety of internal environmentally beneficial actions that contribute to energy savings and sustainable production. These may include the use of low impact vehicles and "Zero km" raw materials.

Table 2 Association rules output (Consequent "Cost Leadership")

Rule no.	LHS	=>	RHS	Support	Confidence	Lift
1	{STAKE_SECO$_2$ = 1, COMP_RISK$_3$ = 1}	=>	{COMPADV_COST = 1}	0.910	0.965	1.5
2	{STAKE_SECO$_2$ = 1}	=>	{COMPADV_COST = 1}	0.930	0.962	1.2
3	{COMP_RISK$_3$ = 1}	=>	{COMPADV_COST = 1}	0.930	0.962	1.2
4	{PRACT_TRANS$_2$ = 1}	=>	{COMPADV_COST = 1}	0.920	0.962	1.1
5	{STAKE_SECO$_2$ = 1, COMP_RISK$_2$ = 1}	=>	{COMPADV_COST = 1}	0.920	0.961	1.1
6	{STAKE_INT_PRIM$_3$ = 1}	=>	{COMPADV_COST = 1}	0.900	0.961	1.1
7	{COMP_RISK$_2$ = 1, COMP_RISK$_3$ = 1}	=>	{COMPADV_COST = 1}	0.907	0.961	1.1

- Rule 7 means that compliance and risk is most likely to be reached when Rivals are an important stakeholder. One-to-one interviews with company managers about this point highlighted that many companies aim at gaining competitiveness, in comparison to their rivals, from various environmental policies and regulation.
- Rule 8 and 9 mean that compliance and risk advantage is most likely to be reached when Environmental auditing programme for suppliers is carried out and Sustainable disclosures are developed.

Analysing the results obtained with "Cost Leadership" as a consequent (Table 2) it is possible to state that:

- In fashion sector, companies that pursued "Lower cost" as competitive advantage aim at obtaining from environmental strategy a great access to capital (rules 1, 3 and 7) and lower cost of capital (rules 5 and 7).
- Moreover, these companies involved in their environmental strategy Environmental NGOs and Financial institutions (rules 1, 2, 5 and 6).
- Many fashion companies are becoming aware of the possible risks that NGOs could present to their business. Moreover, environmental NGOs are recognising the pivotal role of finance in today's economy and are seeking ways to influence where capital flows.

6 Conclusion

In this work an investigation of environmental strategies carried out by in fashion sector has been carried out analyzing a sample of 343 companies.

Several rules have been generated in order to identify relationships among sustainable practices, impacts of environmental strategy and stakeholders typology involved in the strategy.

In the case examined in this work, the intrinsic structure and complexity of the data collected might jeopardize the use of traditional tools for analysis since the variables present the following critical characteristics:

- high dimensionality: a high number of variables is a problem of considerable importance for standard statistical analysis in general;
- interaction and dependency: the relations between the independent variables can be a problem particularly in parametric analyses, which typically adopt the independence hypothesis;
- non-homogeneity and non-linearity: different relationships may exist between the variables in several parts of the measurement space.

As a data mining technique, Association Rules methodology demonstrated to be a powerful alternative to the frequently used traditional parametric techniques.

From this point of view Association Rules can be considered a complementary tool to parametric methods, guiding the researcher towards a more thorough understanding of the data.

References

Agrawal R, Tomasz I, Arun S (1993) Mining association rules between sets of items in large databases. In: Proceedings of the 1993 ACM SIGMOD international conference on management of data. SIGMOD '93. ACM, NY, pp 207–216. doi:10.1145/170035.170072

Baumgartner R, Ebner D (2010) Corporate sustainability strategies: sustainability profiles and maturity levels. Sustain Dev 18(2):76–89

Bevilacqua M, Ciarapica FE, Mazzuto G, Paciarotti C (2014) Environmental analysis of a cotton yarn supply chain. J Clean Prod 82(1):154–165

Blomgren A (2011) Does corporate social responsibility influence profit margins? A case study of executive perceptions. Corp Soc Responsib Environ Manage 18(5):263–274

Caniato F, Caridi M, Crippa L, Moretto A (2012) Environmental sustainability in fashion supply chains: an exploratory case based research. Int J Prod Econ 135(2):659–670

Carmona-Moreno E, Céspedes-Lorente J, De Burgos-Jiménez J (2004) Environmental strategies in Spanish hotels: contextual factors and performance. Serv Ind J 24(3):101–130

Chi T (2011) Building a sustainable supply chain: An analysis of corporate social responsibility (CSR) practices in the Chinese textile and apparel industry. J Text Inst 102:837–848

Choi TM, Chiu C-H, Govindan K, Yue X (2014) Sustainable fashion supply chain management: the European scenario. Eur Manage J 32(5):821–822

Cooperrider D, Fry R (2012) Tied into a single garment of destiny. J Corp Citizsh 45:3–4

Cramer J (1998) Environmental management: from "fit" to "stretch". Bus Strategy Environ 7(3):162–172

De Brito M, Carbone V, Blanquart C (2008) Towards a sustainable fashion retail supply chain in Europe: organisation and performance. Int J Prod Econ 114(2):534–553

Epstein MJ, Roy M (2003) Making the business case for sustainability. J Corp Citizsh 9:79–96

ESSC TCLF (European Sector Skills Council—Textile Clothing Leather Footwear) (2014). Report 2014. http://europeanskillscouncil.t-c-l.eu/en/reports.aspx

Euratex (2014) Annual report. Taking action for the future of the European textile and clothing industry. Euratex, Brussels

Euratex (2015) The EU-28 Textile and clothing industry in the year 2014. http://euratex.eu/library/statistics/key-data/key-data-details/?tx_ttnews%5Btt_news%5D=4846&cHash=a43f98d261ba-88233db94846a8f7e08e

Eurostat (2013a) Manufacture of textiles statistics—NACE Rev 2—Statistics explained. http://epp.eurostat.ec.europa.eu/statistics_explained/index.php/Manufacture_of_textiles_statistics_-_NACE_Rev._2

Eurostat (2013b) Manufacture of wearing apparel statistics—NACE Rev 2—Statistics explained. http://epp.eurostat.ec.europa.eu/statistics_explained/index.php/Manufacture_of_wearing_apparel_statistics_-_NACE_Rev._2

Eurostat (2013c) Manufacture of leather and related products statistics—NACE Rev 2—Statistics explained. http://epp.eurostat.ec.europa.eu/statistics_explained/index.php/Manufacture_of_leather_and_related_products_statistics_-_NACE_Rev._2

Forza C (2002) Survey research in operations management: a process-based perspective. Int J Oper Prod Manage 22(2):152–194

Galbreath J (2009) Building corporate social responsibility into strategy. Eur Bus Rev 21(2):109–127

Henriques I, Sadorsky P (1999) The relationship between environmental commitment and managerial perceptions of stakeholder importance. Acad Manage J 42(1):87–99
Jabbour C (2010) Non-linear pathways of corporate environmental management: a survey of ISO 14001-certified companies in Brazil. J Clean Prod 18(12):1222–1225
Kramer M, Porter ME (2006) Strategy and society: the link between competitive advantage and corporate social responsibility. Harvard Bus Rev 84(12):76–92
Lo CKY, Yeung ACL, Cheng TCE (2012) The impact of environmental management systems on financial performance in fashion and textiles industries. Int J Prod Econ 135(2):561–567
Malhotra MK, Grover V (1998) An assessment of survey research in POM: from constructs to theory. J Oper Manage 16(4):407–425
Murthy V (2012) Integrating corporate sustainability and strategy for business performance. World J Entrep Manage Sustain Dev 8(1):5–17
Na Y, Na DK (2015) Investigating the sustainability of the Korean textile and fashion industry. Int J Clothing Sci Technol 27(1):23–33
Resta B, Stefano D, Pinto R, Bandinelli R, Rinaldi R, Ciarapica FE (2014) Practices for environmental sustainability in the textile, clothing and leather sectors: the italian case. Int J Oper Quant Manage 20(3):101–133
Rhee SK, Lee SY (2003) Dynamic change of corporate environmental strategy: rhetoric and reality. Bus Strategy Environ 12(3):175–190
Salzmann O, Ionescu-Somers A, Steger U (2005) The business case for corporate sustainability: literature review and research options. Eur Manage J 23(1):27–36
Savitz AW, Weber K (2006) The triple bottom line. Jossey-Bass, San Francisco
Smith NC (2003) Corporate social responsibility: whether or how? Calif Manage Rev 45(4):52–76
Sroufe R, Montabon F, Narasimhan R, Wang X (2002) Environmental management practices. Greener Manage Int 40:23–44

Part V
Brand Management and Strategy

Social Media Strategy in the Italian Fashion Industry: A New Model of Analysis

Monica Faraoni, Romeo Bandinelli and Rinaldo Rinaldi

Abstract Research regarding social-media strategy is a rapidly growing field of interest as social media (SM hereinafter) become vital tools for marketing managers to communicate with the consumer who is increasingly eager to share opinions and to be involved in the "brand life". Customer involvement through interaction with a brand is deeply related to the effectiveness of the SM strategy. As a consequence, companies need to carefully define the key elements of their SM strategy and make decisions about goals, the target audience, channel choice, content-planning activities, resource allocation, internal policies, monitoring, and controlling the online activity in order to increase consumer brand awareness and make their SM strategy more effective. The literature provides some models of analysis, but further investigations are still necessary. In particular, it is not clear how certain variables such as the level of brand social engagement or, the company size and the company market segment, can affect the level of importance of the SM strategic dimensions. We hypothesized that the key elements of SM strategy can have variable weights in relation to these variables, and we test our assumptions on fashion industry companies. Analysis was conducted on a total sample of 42 companies, and the results show that there is a significantly different perception about the weight that the single strategic dimension can have. Companies with a high level of social engagement, for example, have a higher perception of the strategic role of the resource allocation, internal policies, and the content definition compared with the perception of lower socially engaged companies; small companies generally do not perceived the importance of monitoring and controlling SM activity highly as compared to large and medium companies, while luxury-brand companies rely more on the strategic role of the target audience dimension, the policy, and the

M. Faraoni · R. Bandinelli (✉) · R. Rinaldi
Department of Economics and Management, University of Florence, Florence, Italy
e-mail: romeo.bandinelli@unifi.it

M. Faraoni
e-mail: monica.faraoni@unifi.it

R. Rinaldi
e-mail: rinaldo.rinaldi@unifi.it

R. Rinaldi and R. Bandinelli (eds.), *Business Models and ICT Technologies for the Fashion Supply Chain*, Lecture Notes in Electrical Engineering 413,
DOI 10.1007/978-3-319-48511-9_21

content-planning activity. Managerial implications about the way the marketing manager can plan a SM strategy are then derived from these results.

Keywords Social media strategy • Fashion brands • Strategic dimensions • Social engagement

1 Social Media Strategy in the Fashion Industry: An Introduction

Social media (SM) has become vital for marketing managers to communicate with consumers, who are increasingly eager to share their opinions and engage with brands. A number of studies have found that, when used effectively, SM can be a valuable communication instrument to enhance the consumer brand engagement (Sashi 2012), which in turn leads to a significant positive impact on long-term customer relationships (Hofmann and Fodor 2010) and brand loyalty (Erdogmus and Cicek 2012). Today's consumer is more powerful than ever. The degree of consumer involvement in brands is closely related to the effectiveness of a company's SM strategy. Specifically, companies need to carefully define the key elements of their SM strategies, including number and types of SM to use, the purposes of the interactions, communication contents, expected results, budget allocations, etc., in order to achieve customer engagement in all facets (Effing 2013). In using SM, managers need to shift the marketing strategic focus from "trying to sell" to "making connections" with consumers (Gordhamer 2009), which implies new definitions of strategic dimensions.

This is especially true for fashion brands that can no longer base their success on secure, regular customers as they could in the past (Kim and Ko 2012). Today, expanding brand image and customer engagement have become critical for positioning in the fashion market (Brun et al. 2008). Consumers of fashion actively interact by means of digital platforms (Boyd Thomas et al. 2007; Rickman and Cosenza 2007), such as online communities of fashion bloggers that are important for creating consumer engagement.

In this context, one of the most effective and nearly cost-free ways to increase companies' appeal is SM (Kim and Ko 2010). Its effectiveness is demonstrated by its growing use among fashion brands (Kim and Ko 2012). Initially, fashion brands showed low commitment in integrating advanced Internet technologies (Okonkwo 2009), as they were not fully convinced of the potential of these channels to attract customers (Kim and Ko 2010); this was especially true for luxury-fashion brands. One of the core features of the Internet is the classless mass media aspect, which seems to contradict the exclusivity and uniqueness of luxury brands (Ng 2014). Today, the luxury industry has gradually recognized the importance of using the internet and SM to enhance a brand's reputation, increase interactions with customers, and stimulate their desire for luxury (Ng 2014). As an example, SM has helped Burberry to revitalize its brand and reposition it to a new, younger market (Phan et al. 2011). Despite this significant attention from the industry, there is still a dearth of scholarly sources from a SM strategy perspective. Structured research is still lacking, and an empirical analysis of these aspects could produce interesting findings.

The present research aims to define and measure the key elements of SM strategy and SM's role for fashion companies.

It represents the first phase of a wider investigation aimed at measuring the relationship between the SM strategic elements used by companies in the fashion industry and the performance marketing KPI's (like the brand reputation index, the level of brand awareness, the brand equity online, etc.) that will be investigated in the second phase of the research.

Consequently, it is intended that the contribution derived from this work will be expanded, once these results have been compared with the marketing performance indexes. From the analysis of the key elements used by companies related to their online marketing results, we will derive important strategic insights both from a theoretical point of view and from a managerial one. Marketing theory can benefit from this contribution because the study will highlight the differences with the communication strategies used in the traditional economy and because it will make possible to build high-performance SM communication models. The measurement of the key elements is a first, very important step of the research that bridges the gap in the literature and enable further progress in marketing theory and practice.

Managers can find important insights for their SM marketing planning activity.

2 Definition of Social Media Strategy

SM is a broad term for online applications, platforms, and media that aim to facilitate interactions, collaboration, and sharing of content. It takes a variety of forms, including weblogs, social blogs, microblogs, wikis, podcasts, pictures, videos, and social networks. SM is defined as "activities, practices and behaviours among communities of people who gather online to share information, knowledge, and opinion using conversational media" (Safko and Brake 2009). Kaplan and Haenlein (2010) expand on this definition by underlining the role of customer content: "SM is a group of Internet based applications that build on the ideological and technological foundations of Web 2.0, and that allow the creation and exchange of user generated content."

SM is a powerful phenomenon that has changed social interactions globally. Therefore, it is crucial for companies today to engage and use social networks in order to be competitive in the market (Burkhalter et al. 2014; Kaplan and Haenlein 2010; Larson and Watson 2011; Othman et al. 2013). SM is changing the way business is conducted by providing low-cost platforms for personal branding (Dutta 2010) and corporate and brand reputation, and, regardless of company size or industry, SM has become a mandatory element of companies' marketing strategies

(Hanna et al. 2011). To operate in this new context, companies need to define their strategies so as to plan and organize activities to create unique and valuable market positions. Given the particular nature of these SM instruments, the concept of strategy should specifically focus on "digital business strategy," defined by Bharadwaj et al. (2013) as "an organizational strategy formulated and executed by leveraging digital resources to create differential value." SM is an important part of a company's digital resources, and researchers consider SM strategy "a goal-directed planning process for creating user generated content, driven by a group of internet applications, to create a unique and valuable competitive position" (Effing and Spil 2016).

Based on a survey of the literature, SM strategy is a rapidly growing field of interest. Contributions underline SM usage for investigating consumer behaviour and monitoring ongoing activities for market research (as in the netnography approach of Kozinets (2002)), and for companies marketing and advertising their offerings to wide online audiences (Curran et al. 2011; Constantinides et al. 2008; Munar 2012; Wilson et al. 2011). Other research focuses on step-by-step strategies for implementing SM in order to determine the best plan for how to introduce and effectively use SM instruments (Hayes et al. 2013; Othman et al. 2013; Rodriguez-Donaire 2012; See-Pui Ng and Chung Wang 2013). Some authors propose specific SM strategies; for example, Wilson et al. (2011) suggest different types of strategies in relation to a company's intention to use SM tools for a specific area such as customer service or for large-scale interactions. Constantinides et al. (2008) distinguish between "active" and "passive" utilization of SM as a marketing tool in the first case or only for monitoring users in the second case.

In a deep analysis of the SM-strategy literature, Effing and Spil (2016) arrives at a list of seven key elements that are often included in authors' contributions on this topic. These elements generally comprise companies' SM strategies, and their correct consideration when approaching SM as a fundamental ingredient in high-quality SM planning. In particular, the SM strategy elements to consider are: setting SM strategy goals in terms of specific expected results; defining the target audience; defining "channel choice" by setting a multichannel versus single-platform policy; deciding on the content planning activities by scheduling posts, including determining the style, content, and types of posts; establishing the levels of resource allocation, both financial and in terms of human working time; setting policies and the best practices for employee to minimize risks; and monitoring and controlling the online customer, brand awareness generated by SM activity. In analyzing the presence of these items in SM strategies in nine cases, Effing and Spil (2016) distinguishes three stages of maturity for SM strategy development: initiation, when the focus is on channel choice and target; diffusion, when policies, goals, and resources are crucial elements; and maturity, when the focus is on monitoring and content. The author introduces the concept of the "social strategy cone" as a framework for companies' SM strategies. Effing and Spil (2016) study is an important contribution to better understanding SM-strategy development because it clarifies the components of SM-strategy plans by assessing their quality.

Following the Effing model, this paper takes into consideration the seven key elements of SM strategy proposed by Effing and goes further in trying to measure the weights of these elements in SM strategies. In other words, while the Eiffel model contribution is that to identify and define the key elements of SM strategy, in our work we measure the importance of these elements for fashion companies. Moreover we also identify another value that they can have when considering some clusters of fashion companies classified by size, segment target, and level of online engagement.

The measurement of the key elements is a contribution to marketing theory and practice as it gives important insight about the strategic priority and, once compared with marketing performance, it makes clear the directions that SM strategy must have for each fashion company.

3 Conceptual Model and Hypothesis Development

In building a conceptual model, as mentioned before, this paper extends the Effing approach by measuring the weight of the seven key elements in specific company clusters of the fashion industry classified using three variables: level of online customer engagement, company size, and company market position. The reason for this analysis is that companies may have different approaches to SM strategy in relation to their online brand reputations, levels of engagement, sizes, and market positions (luxury, accessible luxury, or mass market).

Specifically, the present paper aims to investigate four research questions (RQs), as detailed in the following paragraphs.

RQ1. What is the relevance of the key elements in fashion companies' SM strategies? This question aims to identify fashion companies' perceptions about the key SM elements. Ineffective perceptions can lead to poorly planned SM activity, which could heavily affect online performance.

Perceptions can differ when companies have a great number of followers and the brand has a high level of online awareness and/or customer engagement. Customer engagement is defined as "the level of a customer's cognitive, emotional and behavioural investment in specific brand interactions" (Hollenbeck 2011: p. 565). The relationship between SM and customer engagement is analyzed in the literature in terms of benefits (Brodie et al. 2013; Ramaswamy 2009; Sawhney et al. 2005) or valuable communication (Sashi 2012). Parent et al. (2011) propose a framework they call the six C model that explains how to engage customers through SM; the six key elements are very similar to those of the Effing and Spil (2016) model. The authors consider "companies" in terms of defining the platforms a company wants to utilize for its online presence (similar to channel choice in the Effing model); "content" in terms of accurately selecting subjects to post about; "control" in terms of monitoring; and "communities" in terms of the people who help to spread the word of mouth and thus create online buzz. The final two elements of the framework are the target "customer" and "conversations;" engaging in conversations is key to a successful

SM presence. The relationships that the authors discuss, regarding key elements of company SM behaviour and customer engagement, led to the next research question:

RQ2. Do key elements have different weights in SM strategies for companies with different levels of online, brand customer engagement?

This paper analyzes company size as a discriminant variable in SM strategy. Small businesses can benefit from SM because they can present their products and services to extremely large audiences and still maintain close relationships with them (Lacho and Marinello 2010). Moreover, SM tools are often relatively inexpensive (Kahle and Valette-Florence 2012), which can be a great advantage for small companies with tight marketing budgets. The research question for this aspect is:

RQ3. Does company size matter in SM strategies?

The paper then considers the fashion brand's market position (mass market, accessible luxury, or luxury) as an indicator of a company's SM-strategy behavior. Existing studies on the fashion industry focus mainly on large global luxury fashion brands (e.g. Kontu and Vecchi 2014; Ng 2014) and do not consider small to medium-sized enterprise (SME) brands, including SME luxury brands. The relevant research question here is:

RQ4. Does company market position matter in strategic SM behavior?

4 Empirical Settings and Methodology

The research process is based on an explorative approach that aims to understand the different weights of the key elements of SM strategy used by fashion companies. It is articulated through several phases, as detailed below.

4.1 Sample Definition and the Questionnaire

To investigate the research questions, a self-administrated questionnaire was developed. The online survey was distributed to fashion companies that use SM for their marketing activities, and 42 completed surveys were returned. The questionnaire was structured in five sections: Sect. 1 collected general company information (e.g., dimension); Sect. 2 analyzed companies' SM strategies in both qualitative and quantitative ways, investigating most-used SM tools; Sect. 3 focused on implemented SM activities, their monetary, and time investments and their effectiveness in terms of achieving cognitive, emotional, or behavioral engagement; Sect. 4 investigated the monitoring of SM activities; and Sect. 5 analyzed companies' future SM directions. The sample for this study was particularly comfortable with SM: in addition to their corporate websites, 84% of companies used both a SM institutional page and an e-commerce site; only 11% had a unique presence on

SM. A total of 95% of the surveyed companies used social networks in some way, 81% were present in a content community such as YouTube, 51% had blogs, 3% were present on social news websites, and 5% participated in collaborative projects. Simulation environments such as virtual worlds (Second Life, for example) were not used by any of the companies, showing that these tools are not yet perceived as strategically important, given the limited availability of resources. Moreover, in terms of resource allocation, the data showed that 87% of fashion companies spent up to 25% of their marketing budgets on SM.

4.2 *The Definition of Variables*

RQ1 was tested using questions on the key elements of an effective SM strategy. In particular, for the "Goals" element, this section of the questionnaire divided customer engagement into its components (cognitive, emotional, behavioral) and measured each using specific questions, such as the importance of informing consumers about brand and company events (cognitive engagement), consolidating and strengthening brand attachment (emotional engagement), and sharing opinions on the brand and generating consumer content, website traffic, etc. (behavioral engagement). All these questions formed a specific construct, called D1. D2 considered companies' channel choices by measuring their awareness that different SM channels address different target groups (Dutta 2010). Because each SM channel has its own richness and appropriateness (Kaplan and Haenlein 2010; Smith et al. 2012), this study also measured perceptions of the relationships between SM channels and communication goals on one side and customer engagement on the other. D3 assessed the perceptions of target audiences in the context of SM strategy. Companies need to identify which groups to address with SM, including knowing how different groups use it (Thackeray et al. 2008). D4 considered SM policies and best practices. Organizations need to have basic rules in place to protect their corporate reputations (Mortleman 2011), because SM blurs the boundaries between organizations and the environment. D5 focused on content planning in terms of the timing of campaigns, product promotions, conversational themes, and the arguments that are generally used in SM activity (Klang and Nolin 2011).

D6 measured resource allocation in terms of SM budget and working time. The success of SM strategy depends primarily on the level of resources allocated to it (Dutta 2010). Finally, D7 measured the perceived importance of monitoring and controlling activity by observing what was happening on company SM sites. Standard software tools, together with simple measurement metrics, can help in evaluating online SM activity (Klang and Nolin 2011); examples of such tools include Google Alert, Buzzient, Klout, and Twitalizer (Woodcock et al. 2011) (Fig. 1).

RQ2 was tested using metrics extracted from the Talkwalker platform, which is one of the main social-data, intelligence tools. It processes 500 million posts from 150 million websites every day, monitoring and analyzing all conversations on

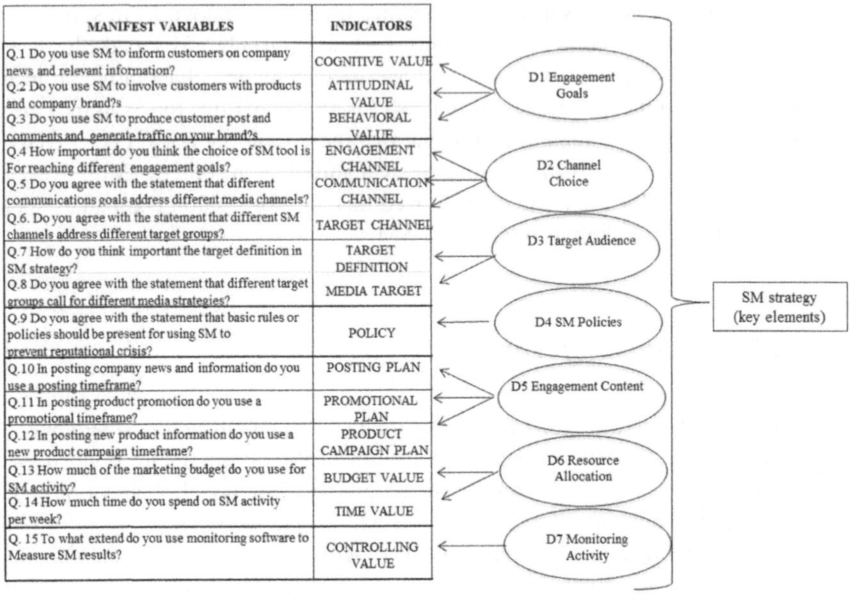

Fig. 1 Manifest variables, indicators and key elements

social networks, news websites, blogs, forums, and other sources in 187 different languages. Talkwalker is used by more than 500 large global brands, including Benetton, KPMG, Merck, and BASF. The research results are aggregated in performance indicators such as the number of "mentions," which measures the number of online mentions and reflects the buzz around a brand–that is, how many people are talking about it in a specific period; "Engagement", which looks at how people interact with brands through posts on SM (likes, shares, favorites) and gives an idea of how viral SM activity is; and "Potential Reach", which measures how many users a company could reach with its activities. Table 1 shows these metrics for each company in the sample.

RQ3 was tested using company size (small, medium, or large) as a discriminant variable measured by the number of employees according to the European Commission's definition of SMEs.

From a descriptive point of view, some evidence emerged initially regarding social networks, blogs, and community presence: whereas large and medium-size companies were present on all SM channels, small businesses did not manage social networks institutionally. This diversity could be related not to different strategic perceptions, but to smaller companies' more limited budgets in terms of both resources and time. This assumption is confirmed by two different results from the survey. First, marketing managers expressed positive opinions regarding the use of these instruments on every dimension, and second, companies invested different levels of dedicated SM resources (up to 25% of the marketing budget for small companies versus more than 25% for larger firms). The same results emerged in the

Table 1 Customer engagement metrics (talkwalker and social mentions)

Company dataset	Mentions	Engagement	Potential reach	Negative	Sentiment neutral	Positive
PEUTEREY	749	1970	233.3	355	281	112
A S. WATSON	8062	180600	68.5 g	3869	2813	1380
CRIMSON	6	9	15.1 m	0	4	2
ROBERTO CAVALLI	4038	32200	24.9 g	211	3144	683
KOCCA srl	238	1400	263.1 m	17	179	42
CONTE OF FLORENCE	24	7	47.8 m	0	17	7
NOMINATION ITALY	59	99	2.8 m	33	21	5
TESSILFORM (PATRIZIA PEPE)	573	4.156	1.2 G	28	428	117
EMILIO PUCCI	1300	188	992.6 M	19	1161	122
MODAE SRL-BRAINTROPHY	51	159	726,20 K	1	28	22
BRACCIALINI	52	1400	1.6 M	1	44	7
PIQUADRO	384	444	685.7 M	11	173	200
LA PERLA	988	1800	863.9 M	125	646	217
MARINA RINALDI—MMFG	85	17	5.6 M	3	0	82
LIU JO	298	1700	403.8 M	15	234	49
DSQUARED2	1717	14400	13.2 G	121	1389	207
CALZEDONIA	5029	72200	2.9 g	358	3352	1319
ANTICA OFFICINA DEL FARM	14	0	286.3 M	0	10	4
SALVATORE FERRAGAMO	1339	2100	2.8 G	82	1105	160
ELISABETH SRL (WALK SAFARI)	3	1	140.1 K	0	0	1
OBERALP	83	39	11.5 M	4	68	11
MABRUN	2	0	22.3 M		1	1
GRISPORT	4	0	0	0	0	4
CELEGHIN	0	0	0	0	0	0
SPIDI SPORT	152	70	55 M	6	99	77
VALENTINO	5175	18800	13.9 G	413	3965	797
OVS Spa	1407	4300	1.5 G	155	744	548
ANTICO SETIFICIO FIORENTINO	0	0	0	0	0	0
LUXOTTICA	1600	30900	4.2 g	231	908	448
LUISAVIAROMACOM	1795	2900	3.2 g	88	1043	664
MANTERO SETA SPA	4	42	20300	0	2	2
PARAH	88.100	773.800	12.6 g	25580	51420	11039
LUISA SPAGNOLI	92	5200	226.9 m	1	62	29
VIONNET	525	4700	3.5 g	28	436	61

analysis of the time dedicated to SM activity (10 h per week for small companies as compared to 20 h for larger firms). These results were the basis for answering RQ3, which considered company size a discriminant variable in the key elements of SM strategy.

RQ4 was tested using company market position, and the sample was divided into two groups (accessible luxury/luxury and mass market) as derived from Saviolo and Testa (2005).

5 Results

Table 2 shows the results of the analysis. Column 1 shows the means for each indicator (I) and each key element (KE) for the full sample in order to answer RQ1. The other columns distinguish companies by the level of customer engagement (RQ2), size (RQ3), and market position (RQ4). For each company classification, the study measured the mean value for each indicator, the aggregated mean value for each KE, and the differences in the key elements per group (ΔKE) within the company class.

For RQ1, the companies reported clear engagement goals (D1), cognitive, attitudinal, and behavioral (aggregated mean KE = 4.58), but the other SM key elements had less weight in strategy formulations.

The "channel choice" (D2), "target audience" (D3), and "content planning" (D5) were rated as relatively equally important (3.53, 3.15, 3.24, respectively), and "monitoring" weighed less (2.51). The "policies and best practices" were not rated as especially important (2.02), and neither was the "resource allocation" (2.00).

For RQ2, the main differences in the key element ratings between companies with high versus low or moderate levels of customer engagement were in "policies" (ΔKE 2.36), "resource allocations" (ΔKE 2.27), and "content planning" (ΔKE 1.65).

Answering RQ3 showed that the main differences in the key SM elements between small and medium–large companies related to "monitoring" (ΔKE 2.62) and "engagement goals" (ΔKE—1.03).

Finally, answering RQ4 revealed consistent differences among luxury and accessible luxury market companies compared with mass-market companies; these differences mainly related to "policies" (ΔKE 1.96) and "content planning" (ΔKE 1.92), whereas the weights were quite different for "target audience" (ΔKE 1.70), "channel choice" (ΔKE 1.15), and "monitoring" (ΔKE 1.18).

Table 2 Indicators and key elements for the full sample (RQ1) and for the sample divided by level of customer engagement (RQ2), company size (RQ3), and market position (RQ4)

MV	M	I	KE	Customer engagement					Company size					Company market position				
				I (HCE)	KE (HCE)	I (LMCE)	KE (LMCE)	ΔKE	I (SC)	KE (SC)	I (MLC)	KE (MLC)	ΔKE	I (LC)	KE (LC)	I (MMC)	KE (MMC)	ΔKE
Q1	4.51			4.83		4.20			4.10		4.93			4.61		4.42		
Q2	4.74	D1	4.58	4.62	4.79	4.87	4.38	0.41	4.70	4.43	4.79	4.73	−0.30	4.90	4.80	4.59	4.36	0.44
Q3	4.49			4.91		4.06			4.50		4.47			4.90		4.07		
Q4	3.82			4.22		3.42			3.20		4.44			4.10		3.54		
Q5	3.56	D2	3.53	3.98	4.07	3.14	2.98	1.09	3.15	3.15	3.97	3.91	−0.76	4.20	4.10	2.92	2.95	1.15
Q6	3.20			4.01		2.39			3.09		3.31			4.00		2.40		
Q7	3.45	D3	3.15	3.80	3.50	3.10	2.80	0.70	3.15	2.83	3.75	3.48	−0.65	4.05	4.00	2.85	2.30	1.70
Q8	2.85			3.20		2.50			2.50		3.20			3.95		1.75		
Q9	2.02	D4	2.02	3.20	3.20	0.84	0.84	2.36	1.80	1.80	2.24	2.24	−0.44	3.00	3.00	1.04	1.04	1.96
Q10	3.52			4.12		2.92			3.22		3.82			3.85		3.19		
Q11	3.21	D5	3.24	4.10	4.07	2.32	2.42	1.65	3.14	3.09	3.28	3.40	−0.31	5.56	4.20	0.86	2.28	1.92
Q12	3.00			3.99		2.01			2.90		3.10			3.20		2.80		
Q13	2.00	D6	2.00	3.15	3.14	0.85	0.87	2.27	1.80	1.65	2.20	2.35	−0.70	2.10	2.30	1.90	1.70	0.60
Q14	2.00			3.12		0.88			1.50		2.50			2.50		1.50		
Q15	2.51	D7	2.51	3.15	3.15	1.87	1.87	1.28	1.20	1.20	3.82	3.82	−2.62	3.10	3.10	1.92	1.92	1.18

Legenda
MV Manifest variables
M Media
I Indicators
KE Key elements of SM strategy
I (HCE) Indicators for companies with a high level of customer engagement
KE (HCE) Key elements of SM strategy of companies with a high level of customer engagement
I (LMCE) Indicators for companies with a low-medium level of customer engagement
KE (LMCE) Key elements of SM strategy of companies with a low-medium level of customer engagement
I (SC) Indicators for small companies
KE (SC) Key elements of SM strategy of small companies
I (LMC) Indicators for medium-large companies
KE (LMC) Key elements of SM strategy of medium-large companies
I (LC) Indicators for companies of luxury market
KE (LC) Key elements of SM strategy of companies of luxury market
I (ALC) Indicators for companies of accessible luxury market
KE (ALC) Key elements of SM strategy of companies of accessible luxury market
I (MMC) Indicators for companies of mass market
KE (MMC) Key elements of SM strategy of companies of mass market

6 Discussion and Conclusions

The results identified various different patterns of fashion companies' behaviors related to SM strategy formulation. The key elements of SM strategy have different weights for the fashion companies in relation to their online customer engagement, size, and market position. Engagement (D1) in all facets (cognitive, attitudinal, and behavioral) is a fundamental aspect of SM strategy for each company, regardless of company online engagement, size, or market position. Setting objectives determines target outcomes, and this should preferably be done before entry into SM; it makes SM strategies more successful because various techniques can be easily evaluated and constantly adapted if necessary. Defining SM goals requires alignment with overall business goals (Keitzmann et al. 2011) in order to ensure full employee commitment on one side and clear definitions of brand identity and brand image on the other (Ranfagni et al. 2016). One specific SM goal is obviously to engage customers in the brand to increase brand reputation and, consequently, brand equity and firm performance. Customer engagement can be pursued from a cognitive direction when companies have the goal of widespread product and corporate news; from an attitudinal direction when company/consumer interaction aims to build positive brand-related effect; and from the behavioral direction when companies aim to stimulate online reactions such as shares, comments, posts, and website searches. Each goal can be better achieved if it is specifically defined at the beginning of SM-strategy formulation, and fashion companies appear to be aware of this need for planning, given the high values for KE indicators in this study.

In terms of channel choice (D2), fashion companies with high online consumer engagement and luxury companies appear to be more aware of the fact that the choice of an appropriate SM platform (Facebook, Twitter, and YouTube were the most frequently named sites) can affect overall communication. Selecting a SM platform should be based not only on companies' goals (advertisement, product promotion, building brand reputation) but also on target audience, platform preference, and SM usage. Researchers (Lehmkuhl et al. 2013) argue that it is crucial to have a proper Web presence on various SM platforms and to integrate the various platforms to form an entire SM-communication infrastructure. Further, Hanna et al. (2011) suggest that, as companies develop SM strategies, platforms such as YouTube, Facebook, and Twitter are too often treated as stand-alone elements rather than as parts of integrated systems. Companies need to treat SM as an ecosystem of related elements that involve both digital and traditional media. The attention to SM channels by fashion companies with high online engagement and by luxury fashion companies can be explained by the fact that, because brands are followed by users across a variety of online platforms, brand reputations are very vulnerable. This is also why the D3 variable (the target audience) was weighted more heavily by luxury companies than by the others; luxury companies need to pay attention to gain significant reach and attract their target audiences. Determining the right target audience can result from either internal considerations about which group is most likely to purchase certain products or services (segmentation) or by consumer research (Thackeray et al. 2012). Luxury companies reach out to and address key opinion leaders, and these influencers are often the only way to gain and/or maintain brand awareness and lead other users to talk about the brands. Targeting the active participants (Kang et al. 2014) on SM helps to improve brand reputation and popularity because consumers who engage could become brand advocates and create buzz online.

Fashion companies with high levels of customer engagement and luxury companies both rated policies (D4), content planning (D5), and monitoring more highly than other key elements. These companies are more concerned about the risks related to SM (See-Pui Ng and Chung Wang 2013; Vaast and Kaganer 2013), and they attempt to minimize risk by fostering employer attention to SM policies. These policies should ensure the appropriate use of SM (Thackeray et al. 2012) and provide guidelines for employees' usage (Constantinides et al. 2008), directing what they can and cannot post on company websites; this can prevent misguided situations. SM policies are also useful in determining a company's degree of information disclosure (Burkhalter et al. 2014). In this sense, employee training on SM policies can avoid reputational crises and the disclosure of sensitive data. A SM content plan implies creating a schedule for corporate posts (Barnes 2014) that includes the type of content, frequency, and subjects of posts. The regularity of posting (Shen and Bissell 2013) and content variety (Barnes 2014) are important in attracting followers. Moreover, the posts' style, language, and attitudes are crucial for companies that base their strength on their online reputations. In order to foster engagement, posting activity needs to be oriented more towards interactions than promoting products to create a sense of membership among users (Shen and Bissell 2013); this feeling of membership is the basis of brand reputation and brand engagement.

Attention to SM monitoring is higher for fashion companies and larger companies because of their concerns for their brand reputations. Monitoring data is fundamental for marketing research (Malthouse et al. 2013), specifically for minimizing the potential impact of negative posts or user conversations. Moreover, the number of likes, posts, or retweets on companies' SM sites can provide an overview of the value of their communication activities. Comparing these numbers with those of competitors might provide a measure of overall SM performance. Monitoring can be accomplished with available monitoring tools (Constantinides et al. 2008; Jansen et al. 2009). These tools can help firms better understand their acceptance among users and the effectiveness of their actions. Finally, in terms of resource allocation, only fashion companies with high customer engagement appeared to be aware of the need to allocate sufficient financial and time resources to SM activity; employees need to be trained and specifically allocated to this task. In general, there was a tendency to use existing marketing staff members to play this role without specific division of labor. The low attention paid to this variable by the companies in the sample appears to confirm that companies are not completely ready to operate within SM.

7 Managerial Implications and Research Limitations

The present study examines the relevance of the key elements in SM strategy formulation. The study findings support the following conclusions.

First, SM managers of fashion companies need to carefully define SM plans before initiating social website activity in terms of clearly defined engagement goals, in addition to precise identification of company target audiences and the appropriate media platforms in which to operate. Engagement content must be scheduled, not improvised. Employees must be properly trained in order to avoid any reputational damage. The budget allocation has to be determined as a part of the total marketing budget, and monitoring will give accurate information on SM results. All fashion companies appeared to be aware of this, although they weighed various elements (budget allocation and policies, for example) differently.

Second, fashion companies with high levels of online brand awareness and/or customer engagement and luxury fashion companies had higher perceptions of the importance of the key elements of SM strategy that were analyzed in this study. This result can be explained by the different levels of attention marketing managers pay to fashion brand value. On one side, managers perceive greater risks in SM because online reputational crises could lead to uncontrollable brand value damage; this risk encourages SM managers to more carefully plan their companies' SM activities. From the other side, however, marketing managers are aware that SM activity facilitates brand management by providing a type of open window into their customers that traditional marketing media do not usually provide. SM platforms offer venues for customers to engage in sincere and friendly communication with brands, and fashion companies can orient customer preferences, strengthen relationships, build brand reputations, and create purchase intentions. Proper SM planning, aimed at the right target audience and using the right media platform, may enable the achievement of multiple brand management objectives without high resource allocation.

Finally, company size appeared to be nearly irrelevant for the perceptions of the key elements of SM strategy. Large companies' perceptions were slightly higher than those of SMEs; the only important difference was that mainly larger companies engaged in monitoring activity.

In conclusion, the main contribution of this study is that it provides a conceptual framework by analyzing the key elements of SM strategy as used by fashion companies. This information can be helpful for marketing managers who are developing SM strategies because they can identify the elements they need to focus on.

The primary limitation of the study is its small sample size; it would be appropriate to validate the results obtained with a larger sample. In addition, studies on companies in other sectors would be valuable.

The second phase of this research will investigate the relationships between the key elements of SM strategy and SM performance, such as brand reputation and customer engagement. It would be of great interest to learn, for example, whether customer engagement is affected more by multichannel or single-channel SM strategies, by the level of budget allocation, or by a particular target definition.

References

Barnes I (2014) Twitter in special libraries: a distributed social media strategy. Public Serv Q 10 (1):62–65

Bharadwaj A, El Sawy OA, Pavlou PA, Venkatraman NV (2013) Digital business strategy: toward a next generation of insights. Mis Quarterly 37(2):471–482

Boyd Thomas J, Okleshen PC, Tolson H (2007) An exploratory investigation of the virtual community MySpace.com: What are consumers saying about fashion?. J Fash Market Manage: An Inter J 11(4):587–603

Brodie RJ, Ilic A, Juric B, Hollebeek L (2013) Consumer engagement in a virtual brand community: an exploratory analysis. J Bus Res 66(1):105–114

Brun A, Caniato F, Caridi M, Castelli C, Miragliotta G, Ronchi S, Sianesi A, Spina G (2008) Logistics and supply chain management in luxury fashion retail: empirical investigation of Italian firms. Int J Prod Econ 114(2):554–570

Burkhalter JN, Wood NT, Tryce SA (2014) Clear, conspicuous, and concise: disclosures and Twitter word-of-mouth. Bus Horiz 57(3):319–328

Constantinides E, Lorenzo-Romero C, Gomez Boria MA (2008) Social media: a new frontier for retailers? Eur Retail Res 22:1–28

Curran K, Graham S, Temple C (2011) Advertising on Facebook. Inter J E-Business Develop 1(1) 26–33

Dutta S (2010) What's your personal social media strategy? Harvard Bus Rev 88(11):127–130

Effing R (2013) Social media strategy design. In: The 2nd scientific conference, information science in an age of change, pp 1–10

Effing R, Ton AM Spil (2016) "The social strategy cone: towards a framework for evaluating social media strategies. Int J Inf Manage 36(1):1–8

Erdogmus IE, Cicek ML (2012) The impact of social media marketing on brand loyalty. Soc Behav Sci 58:1353–1360

Gordhamer S (2009) 4 ways social media is changing business. http://Mashable.com/2009/09/22/Social-Media-Business/. Accessed 29 Dec 2014

Hayes S, Page SJ, Buhalis D (2013) Social media as a destination marketing tool: its use by national tourism organisations. Current Issues Tourism16(3):211–239

Hoffman D, Fodor M (2010) Can you measure the ROI of your social media marketing? MIT Sloan Manag Rev 52(1):40–50

Hanna R, Rohm A, Crittenden VL (2011) We're all connected: the power of the social media ecosystem. Bus Horiz 54(3):265–273

Hollenbeck L (2011) Exploring customer brand engagement: definition and themes. J Strat Market 19(7):555–573

Jansen BJ, Zhang M, Sobel K, Chowdury A (2009) Twitter power: tweets as electronic word of mouth. J Am Soc Inform Sci Technol 60(11):2169–2188

Kang J, Liang T, Fiore AM (2014) Enhancing consumer–brand relationships on restaurant Facebook fan pages: maximizing consumer benefits and increasing active participation. Int J Hosp Manag 36:145–155

Kaplan AM, Haenlein M (2010) Users of the world, unite! the challenges and opportunities of social media. Business Horizons (53):59–68

Kietzmann JH et al (2011) Social media? Get serious! Understanding the functional building blocks of social media. Bus Horiz 54(3):241–251

Kahle Lynn R, Valette-Florence P (2012) Marketplace lifestyles in an age of social media: theory and method. M. E. Sharpe, Armonk, NY
Kim AJ, Ko E (2010) Impacts of luxury fashion brand's social media marketing on customer relationship and purchase intention. J Global Fashion Mark 1(3):164–171
Kim AJ, Ko E (2012) Do social media marketing activities enhance customer equity? An empirical study of luxury fashion brand. J Bus Res 65(10):1480–1486
Klang M, Nolin J (2011) Disciplining social media: An analysis of social media policies in 26 Swedish municipalities. First Monday 16(8)
Kontu H, Vecchi A (2014) Why all that noise? Assessing the strategic value of social media for fashion brands. J Global Fashion Mark (Special Issue "Luxury Brands and Social Media Marketing) 5(3): 236–251
Kozinets RV (2002) The field behind the screen: using netnography for marketing research in online communities. J Mark Res 39(1):61–72
Lacho K, Marinello C (2010) How small business owners can use social networking to promote their business. Entrep Exec 15:127–134
Larson K, Watson R (2011) The value of social media: Toward measuring social media strategies. Thirty Second International Conference on Information Systems, Shanghai (December 7). ICIS Proceedings Paper 10 http://aisel.aisnet.org/icis2011/proceedings/onlinecommunity/101-18
Lehmkuhl T, Baumol U, Jung R (2013) Towards a maturity model for the adoption of social media as a means of organizational innovation. System sciences (HICSS), 2013 46th Hawaii international conference IEEE
Malthouse EC, Haenlein M, Skiera B, Wege E, Zhang M (2013) Managing customer relationships in the social media era: introducing the social CRM house. J Interact Mark 27(4):270–280
Mortleman J (2011) Social media strategies. Comput Fraud Secur 2011(5):8–11
Munar (2012) Social media strategies and destination management. Scandinavian J Hospitality Tourism 12(2):101–12
Ng M (2014) Social media and luxury fashion brands in China: the case of coach. J Global Fashion Mark 5(3):251–265
Okonkwo U (2009) Sustaining the luxury brand on the internet. J Brand Manag 16(5):302–310
Othman I, Bidin A, Hussain H (2013) Facebook marketing strategy for small business in malaysia. In Informatics and Creative Multimedia (ICICM), 2013 International Conference on IEEE (pp. 236–241)
Parent M, Plangger K, Bal A (2011) The new WTP: Willingness to participate. Business horizons 54(3):219–229
Phan M, Thomas R, Heine K (2011) Social media and luxury brand management: The case of Burberry. J Global Fashion Marketing 2(4):213–222
Ramaswamy V (2009) Leading the transformation to co-creation of value. Strategy & Leadership 37(2):32–37
Ranfagni S, Crawford Camiciottoli B, Faraoni M (2016) How to measure alignment in perceptions of brand personality within online communities: Interdisciplinary Insights. J Interactive Market 35(3):70–85
Rickman AT, Cosenza RM (2007).The changing digital dynamics of multichannel marketing: The feasibility of the weblog: text mining approach for fast fashion trending. J Fash Market Manage: Inter J 11(4):604–621
Rodriguez-Donaire (2012) Changing business model dynamics by means of social media. Proceedings of the 2012 IEEE ICMIT 370-377.
Safko L, Brake DK (2009) The social media bible: tactics. Tools & for Business Success, Hoboken, New Jersey.
Sashi CM (2012) Customer engagement, buyer-seller relationships, and social media. Manag Decis 50(2):253–272
Saviolo S, Testa S (2005) Le imprese del sistema moda. Il management al servizio della creatività, Etas Libri
Sawhney M, Verona G, Prandelli E (2005) Collaborating to create: The Internet as a platform for customer engagement in product innovation. J Interactive Market 19(4):4–17

See-Pui Ng C, Chung Wang WY (2013) Best practices in managing social media for business. In: Thirty fourth international conference on information systems, Milan, pp 1–11

Shen B, Bissell K (2013) Social media, social me: a content analysis of beauty companies' use of Facebook in marketing and branding. J Promot Manag 19(5):629–651

Smith A, Fisher E, Yongjian C (2012) How does brand-related user-generated content differ across YouTube, Facebook, and Twitter? J Interact Mark 26(2):102–113

Thackeray R, Neiger BL, Hanson CL, McKinzie JF (2008) Enhancing promotional strategies within social marketing programs: use of web 2.0 social media. Health Promot Pract 9(4):338–343

Thackeray R, Neiger Brad L, Keller H (2012) Integrating social media and social marketing a four-step process. Health Promot Pract 13(2):165–168

Vaast E, Kaganer E (2013) Social media affordances and governance in the workplace: an examination of organizational policies. J Comput-Mediated Commun 19:78–101

Wilson HJ, Guinan PJ, Parise S, Weinberg BD (2011) What's your social media strategy? Harv Bus Rev 89(7/8): 23–25

Woodcock N, Green A, Starkey M (2011) Social CRM as a business strategy. J Database Mark Custom Strat Man

Fashion Supply Chains and Social Media: Examining the Potential of Data Analysis of Social-Media Texts for Decision Making-Processes in Fashion Supply Chains

Samaneh Beheshti-Kashi, Karl Hribernik, Johannes Lützenberger, Dena Arabsolgar and Klaus-Dieter Thoben

Abstract Fashion companies often face challenges in meeting the demand of consumers since often production plans have to be placed before exact knowledge of the demand is available. Since the industry is a highly consumer- and trend-oriented industry, predicting the customers demand is crucial for the company's success. Nowadays, these customers have been empowered through the Web 2.0 and are able to publish opinions and experiences on various social-media applications. At the same time, these consumers are members of the fashion supply chain. This paper considers a typical fashion supply chain and focusses on the role of the buyer, whose function resides with the retailer. The buyer plays a crucial role since she or he is responsible for the trend monitoring and selection of future fashion collections. The objective of this paper is to examine if social-media text data shared by means of fashion blogs contains color information and if these color comments correspond to real-world customer demand. For this purpose, 232 blog

---

S. Beheshti-Kashi (✉)
International Graduate School for Dynamics in Logistics (IGS),
University of Bremen, Hochschulring 20, 28359 Bremen, Germany
e-mail: bek@biba.uni-bremen.de

K. Hribernik · J. Lützenberger
BIBA - Bremer Institut für Produktion und Logistik GmbH at the University of Bremen, Hochschulring 20, 28359 Bremen, Germany
e-mail: hri@biba.uni-bremen.de

J. Lützenberger
e-mail: lue@biba.uni-bremen.de

D. Arabsolgar
DenaMilano by Laura Mandelli, via Porpora 113, 20131 Milan, Italy
e-mail: dena@denacashmere.it

K.-D. Thoben
Faculty of Production Engineering, University of Bremen,
Badgasteiner Straße 1, 28359 Bremen, Germany
e-mail: tho@biba.uni-bremen.de

R. Rinaldi and R. Bandinelli (eds.), *Business Models and ICT Technologies for the Fashion Supply Chain*, Lecture Notes in Electrical Engineering 413,
DOI 10.1007/978-3-319-48511-9_22

posts were collected, analyzed, and compared to qualitative information on colors provided by a real-world clothing company. The analysis shows that it is indeed possible to discover color information from fashion blogs. Moreover, it revealed that the information identified in the blogs correspond with real-world customer demand.

Keywords Social media • Fashion supply chains • Fashion blogs • Fashion buying • Social media text data analysis

1 Introduction

Fashion markets are characterized by some highly specific peculiarities, such as an extremely high number of product variables in terms of style/material/color and deep customization of products. In addition, the point-of-sale (POS) has a huge impact on the success of products. Since purchasing fashion articles is often an impulse decision by the customer, the availability of the products is a crucial aspect. Therefore, the success of a product is highly dependent on the customers' purchasing behavior. However, these customers are often influenced by various factors such as their current life situation and social factors but also by current trends. Consequently, fashion is highly consumer and trend oriented. Hence, the demand is highly volatile and generates uncertainties for the companies involved. In order to achieve a more precise understanding of the customer needs and to improve the customer experience, feedback from the customer to fashion companies is needed. The EU H2020 Factories of the Future project FALCON addresses these challenges by collecting customer feedback from the Internet, for instance from social media, analyzing this data, and finally providing the analyzed data to the corresponding decision makers. FALCON focusses on the collection and analysis of customer comments, searches, and propositions. Based on the results, fashion companies will be able to conduct, for instance, trend analysis and understand which garments the customers prefer. In this way, decisions such as selection of future fashion collections might be supported and improved. These decisions might impact processes in the whole supply chain. Hence, in addition to social-media analysis, this paper will also give a short introduction to fashion supply chains. In particular, the role of the buyer as a function within the retail segment will be a major focus. She/he is responsible for a range of activities. These include, amongt others, the trend analysis that is often the foundation of decisions for future collections. In order to make fundamental decisions, information regarding colors, prints, or cut are required. This paper focusses on the color feature. The objective is to examine if social-media text data contains color information and if these color comments correspond to real-world customer demand.

The rest of the paper is structured as follows: Sect. 2 presents, firstly, background information on fashion supply chains and focusses on the role of the buyer. Secondly, it gives an overview of social media, particularly fashion blogs, and points out the role of fashion bloggers as influencers and their relevance to the fashion industry. Section 3 summarizes some research conducted in social-media analysis. The methodology is introduced in Sect. 4 which is followed by the final Sect. 5 on data and results. The paper concludes with some discussion and further research aspects.

2 Background

The first part of the chapter will focus on fashion supply chains, introduce the main activities of the buyer, and focus trend monitoring. The second part will give a short overview of social media in general, and focus particularly on fashion blogs.

2.1 Fashion Supply Chains and Decision Making Processes

Figure 1 illustrates a typical fashion supply chain. It shows the various cycles: the procurement, manufacturing, replenishment, customer, and consumption cycles, alongside the main stakeholders of the chain: supplier, manufacturer, distributor, retailer, customer, and consumer.

Each of the stakeholders has a different function, processes, and decisions that have to be made in order to keep the chain running successfully. Accordingly, each of the cycles consists in itself of a number of processes, stakeholder, roles, and functions. For these decisions, the different stakeholders require a particular type of needed information. This paper takes the perspective of the retailer. Figure 2 shows

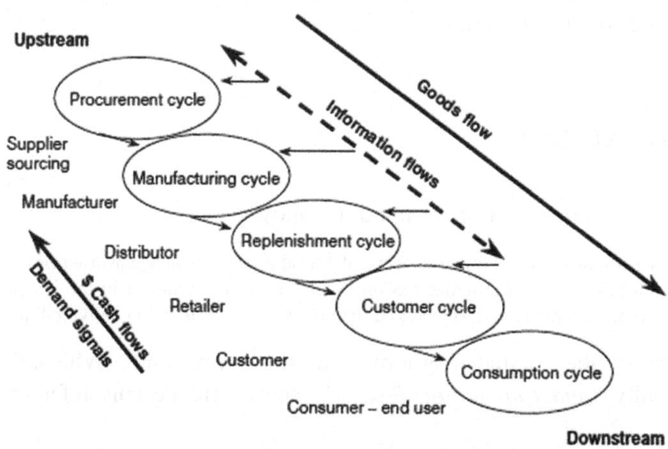

Fig. 1 Supply chain processes (Hines 2014)

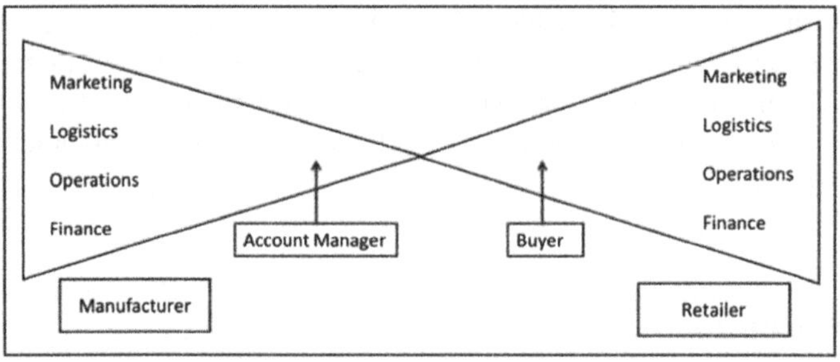

Fig. 2 Interface between manufacturer and retailer (Fiddis 1997)

the interface between manufacturer and retailer. The account manager on the side of the manufacturer is in charge of communication with the buyer or the buying team on the side of the retailer.

The fashion buyer has a crucial function and his/her decisions impact the success of the company (Wong 2013). Therefore, we will focus on this function in the following section. A typical buyer has various responsibilities in a company. He manages the collections and is responsible for negotiations with suppliers, negotiating, among other factors, about prices and deliveries. Furthermore, she/he identifies target groups and examines the buying behavior of the customers. She/he has to predict the sales of the products. Moreover, she/he has to continuously monitor upcoming trends in order react properly to market changes (Jackson and Shaw 2001). For the purpose of this paper, we will describe in more depth the task of trend monitoring. A buyer monitors and evaluates various products and fashion trends in order to decide which products to include in the assortment for a particular season and to be able to react in a timely manner to market changes. For this purpose, the buyer requires information, in particular, of future fashion trends. Therefore, it is required to look deeper into the topic of fashion trends, which will follow in the following section.

3 Trend Analysis

Jackson (2014) defines a fashion trend as follows:

> The term fashion trend refers to aspects of the appearances and construction of fashion products that relate to a particular season. Such trends are manifest in the appearance of fashion products, which are designed and manufactured prior to being delivered in a season.

In addition, they define long-term and short-term trends. Whereas long-term trends usually "*underpin future designs*", short term designs refer to a certain season.

However, what exactly do these trends manifest? Jackson (2014) reports that the features color, fabric, print, silhouette, styling detail, and trim "*can be manipulated to reflect changing fashion"*. Each of the features themselves can stand for a trend, but often a fashion trend consists of a combination of various features. Along the apparel supply chain, the relevance of a fashion trend will be different for the different stakeholders. The information needs will vary according to their position and functions along the chain. However, ideally, a buyer requires information on all the features (Jackson 2014). In order to obtain this information, traditionally, buyers visit fashion shows, fairs, or analyze sales data, among other tactics. Beheshti-Kashi and Thoben (2014) suggested integrating social-media text data for the prediction of future fashion trends. The current paper will have a look at fashion blogs and analyze if they contain information on colors. In addition, it will examine if these colors correspond with real-world customer demand. Therefore, the next section will give an overview on social media and fashion blogs, in particular.

3.1 Social Media

With the advent of the Web 2.0, a huge number of different tools have emerged. These include social networks, blogs, microblogging services, wiki, video-, and photo-share platforms. Each of these tools usually focuses on one type of content. However, mixed usage of the different formats is often practiced. In the literature, we can find various different definitions of social media (Kaplan and Haenlein 2010; Boyd and Ellison 2007). A typical attribute of social-media applications is the lack of filtering of the information. Any ordinary user with access to the web may publish information. Furthermore, with the increased smart phone and mobile web usage, engagement with social-media applications becomes independent from place and time (Eimeren 2013).

About these various applications, a variety of topical aspects are published and discussed. From electronics, travel, beauty, sports, celebrities, politics to fashion, pictures, videos, or comments can be found on the various applications. Focusing on the blogs, we can monitor a huge diversity of topics. One of the first classifications of blogs was conducted by Zerfaß and Bogosyan (2007). They identified, among others, private blogs, corporate blogs, media blogs, and war blogs. One type of blog that has attracted great attention over recent years, is the fashion blog. The following section gives an overview of fashion blogs, their development, and their relevance for the fashion industry.

3.1.1 Fashion Blogs

Fashion blogs have attracted great attention over recent years. They are often authored by young women (but not exclusively) who often write about lifestyle, fashion, beauty, or travelling topics. The following two sections give an overview of a typical blog structure, the fashion bloggers' role as influencers, and their importance for the fashion industry.

3.1.2 Structure of Fashion Blogs

Fashion blogs can be divided into different categories, according to their focus. The main differentiating point is whether the blogger wants to put her/himself into the focus or other people. The first group of blogs usually presents themselves in and their daily life. Accordingly, they are called outfit blogs. The second group focuses on capturing the styles on the streets. Therefore, they are also called street style blogs. This paper focusses on outfit blogs, since text data is often included alongside pictures of the outfits. In outfit blogs, the bloggers present themselves, their daily lives, and routines. Often, they match together outfits and present them on their blogs under the categories my-looks/looks/outfits. Mostly, pictures from various perspectives, descriptions of the outfits, mentions of the different products, information about these articles, brand names, and links to shops where the item might be purchased are published. If the product is older, sold out, or handmade, similar items are suggested for purchased.

3.1.3 Importance for the Fashion Industry

Bloggers are often considered as opinion leaders or influencers (Uzunoğlu and Misci Kip 2014). Usually, they have broad outreach to their readers and followers. Their impact on them is often explained by the phenomenon that the influence of interpersonal communication on an individual's behavior is larger than the mass medias' impact (Weimann 1994). The fashion industry has realized this influence and, in 2006, the first fashion bloggers were invited to report from the New York Fashion Week. Nowadays, fashion bloggers have attained huge importance and are mostly a fixture of international fashion shows. In addition, companies have adapted strategies to involve the bloggers in reaching out to their target groups. These cooperation strategies involve sending out products, organizing events for the bloggers, or inviting them to events such as shop openings. However, before engaging the bloggers, one must identify the relevant influencers. Peng and Sun (2012) reported on the tools that academic researchers consider useful, i.e., various methods for identifying the key influencers. Most of them are using network-structure metrics, such as the number of followers or page rank. Further metrics are the frequency of posts or the number of likes. However, they report that these metrics are not sufficient, since they do not consider the specific topical expertise and the relevant time window. Consequently, they propose a method that includes these two elements, and they published their approach for identifying key influencers more accurately as a U.S. patent. Having identified the adequate influencers, companies may have larger outreach to their target groups through the engagement of fashion bloggers.

For the purpose of this paper, the discussion of the bloggers influence, the detailed strategies regarding influencer cooperation, and actual identification metrics for the most influential bloggers are not treated further and will remain for future study.

3.1.4 Social-Media Text Analysis

Typically, social-media content takes the form of unstructured text data. In order to extract and analyze information from this data, one needs to apply text mining methods. However, before the actual text mining is possible, firstly, natural language processing (NLP) methods need to be conducted in order to transform the data into a structured form. With the emergence of the social-media tools, research on different aspects on social media also have increased. One research stream is the exploration of the predictive value of User-Generated-Content (UGC). Following this stream, diverse authors examine potential relationships between online discourse and real-world outcomes. However, most of this research is conducted with Twitter data. In addition, there is hardly any analysis done in the application field of fashion. A short overview of the predictive value of UGC is provided by Beheshti-Kashi et al. (2015). A second famous research stream is subjectivity analysis, which includes opinion mining and sentiment analysis (Khan et al. 2014).

4 Methodology

As mentioned in the previous section, often opinion mining is conducted on social-media text data in order to extract sentiments from the text. However, we claim that, before extracting any kind of sentiments, it is required to check if fashion blogs contain the required information on the various features. Accordingly, the following two questions shared the focus:

1. Do fashion blogs contain information on colors?
2. Do the occurrences of the colors comments and the real-world customer demand correspond?

In order to answer these questions, two different types of information are required. First, it is needed to have textual data from fashion blogs in order to examine the occurrence of potentially relevant information published on them. Second, in order to analyze potential relationships with real-world processes, it is required to have information on sales or, in general, on customer demand information of a real-world fashion company. An Italian clothing company provided us with information, for instance, on various shapes, materials, or colors for the time period October 2014 to January 2015. This information was available in a rather qualitative form. However, it was applied in order to determine accurate answers the two questions.

In order to compare both data, it was crucial to collect blog data exactly from the given time period. Since the real world information was from an Italian company with stores located in Italy, it was accordingly required to use Italian fashion blogs. However, the language of the collected text data is English. For the analysis, we considered a color list that was provided by the company and searched for the mentioned colors in the blog posts. The color list contained the in-demand and purchased colors for the given time period. In addition to the colors on the list, the 'color' white was added. The primary goal of this examination was to find out if we can extract information on colors from the blog posts. Figure 4 shows the distribution of the mentioned colors. Frequency analyses have been conducted, which identified 290 color occurrences in total.

5 Data and Results

Table 1 shows some characteristics of the blog corpus.

Figure 3 illustrates the post distribution of the five blogs split into the four months. During the time frame October 2014 to January 2015, 232 posts (articles) were published on the five blogs under the categories *outfits*, *my looks,* and *looks*. The other categories, such as beauty, travel and lifestyle, are neglected for the purpose of this paper. With 67 articles, the most posts were published in October, followed by November (62), December (52), and January (51).

Surprisingly, not only dark colors such as black, brown, and grey predominate the corpus. Normally, these colors are expected, since they are typical autumn/winter colors. But, the colors red, white, and pink seemed to have also been relevant during the time period. In particular, red occurred 77 (see Fig. 4) times in the corpus and seemed to be a leading color. Also pink, which is often associated with the seasons spring/summer, is mentioned relatively often compared to, for instance, brown which is a typical autumn/winter color. Comparing this result to the real-world information, we can state that, in the case of red and pink, our finding

Table 1 Corpus statistics

Number of blogs	5
Number of posts	232
Language	English
Time frame (period)	October 2014–January 2015
Number of colors	290

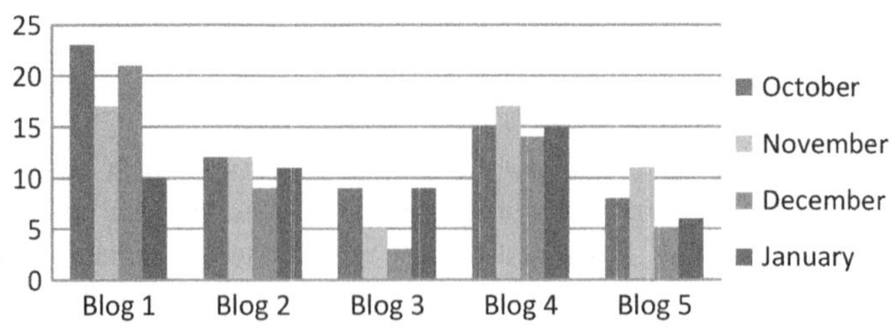

Fig. 3 Post distribution

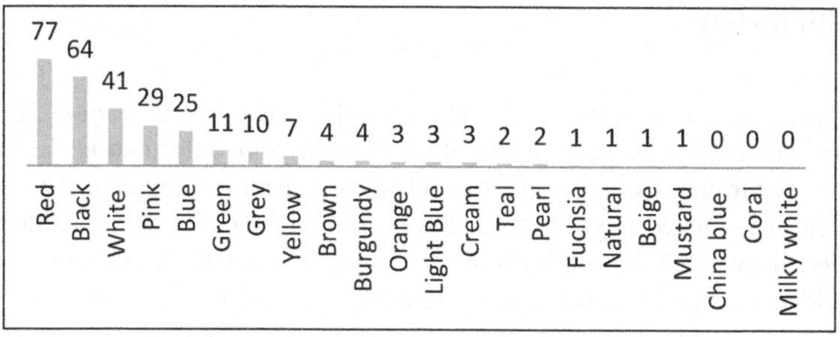

Fig. 4 Color distribution

corresponds to the actual data. However, in the case of white, the two data sets do not correspond. We can find white within the blog posts but not in real-world data. After discussing this issue with the clothing company, we have learned that, in knitting, the color white is hardly used, since the yarn has to be decoloured. In these cases, often the names cream, milky white, or light natural are used instead of white. This case has revealed a further challenge for the research, i.e., the diverse names of colors.

Furthermore, it has shown that it is required to also include experts' feedback into the analysis. The colors burgundy, yellow, or mustard belong to the in-demand colours, although, in the blog posts they do not appear often. However, one interesting observation is that burgundy and yellow have their most mentions in October (see Fig. 5). This corresponds exactly with the provided information.

Figure 5 illustrates the occurrences of the different colors on a time axis. This illustration shows the development of the different colours over the four months. For illustration purposes we deleted the colors that were mentioned three or fewer times. We can observe that the trend for colors is characterized by up and downs over the time period.

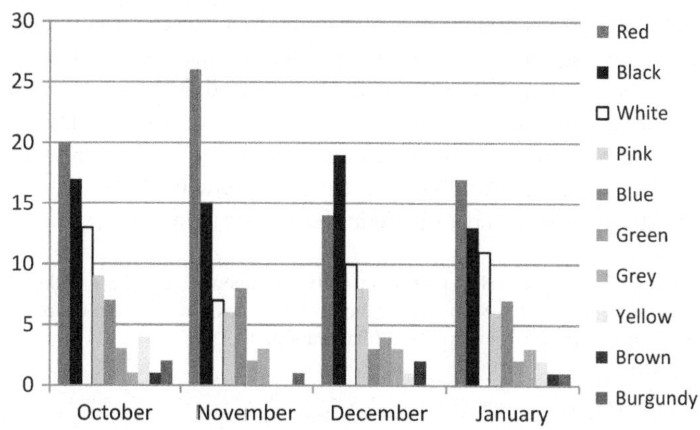

Fig. 5 Monthly color distribution (Color figure online)

6 Conclusion

In this paper, we have analyzed 232 blog posts from five Italian fashion blogs. The objective was first to examine if the blogs contain color information; secondly, if the color occurrences correspond to real-world customer demand. The analysis shows that it is indeed possible to discover color information from fashion blogs. Moreover, it revealed that the information identified in the blogs correspond with real-world information. In particular, the occurrence of the colors red and pink in both datasets supported this the answer to this question. However, we suggest that an analysis of textual color information by quantity (as in this paper) can only be a guideline for the buyer or, in general, the stakeholders in the supply chain. More in-depth analysis is required to catch the popularity of colors in a certain time frame. In particular, the analysis might be enhanced through consideration of an additional semantic layer over the colors that, for instance, reveals the relationship of the colors to each other, or to additional features such as prints. For this purpose, co-occurrence analyses have to be conducted. In addition to further text analysis, an additional layer of computer-vision analysis might be added to match text with photo colors in order to overcome the inconsistency in the naming the colors.

What implications do these results have for the supply chain? For a buyer, it means that, in addition to the traditional ways of monitoring or catching fashion trends, such as visiting fashion shows or fairs, it is worth considering fashion blogs in order to extract color information. To what extent it is possible to discover information on other features such as prints or silhouettes remains for future work.

Given the case that all information that a buyer requires for reasonable decision making might be extracted from fashion blogs, potential implications for the whole supply chain have to be discussed more deeply. For instance, if the buyer will be able to make decisions in a timelier manner due to the consideration of the fashion blog analysis, would other stakeholders such as manufacturers or suppliers be impacted by this development. And if so, how are they impacted? Would this fact require changes in their (manufacturers/suppliers) processes? In order to make fundamental assumptions about these questions, more research is required.

Furthermore, it might be useful to look deeper into the published text data, some weeks or months before and after the actual time period. Moreover, we assume that, in order to make fundamental assumptions, one needs to examine a far larger corpus. Therefore, we will generate a larger corpus, repeat the analysis, and compare the results. In addition to color, further features will be considered.

As the next step, we intend to include the feature print.

Acknowledgements This research is part of the H2020 Framework Program, in the context of the FALCON project (Grant No. 636868). The authors wish to acknowledge the Commission and all the FALCON project partners for a fruitful collaboration.

References

Beheshti-Kashi S, Karimi HR, Thoben K-D, Lütjen M, Teucke M (2015) A survey on retail sales forecasting and prediction in fashion markets. Syst Sci Control Eng 3(1):154–161. doi:10.1080/21642583.2014.999389

Beheshti-Kashi S, Thoben KD (2014) The usage of social media text data for the demand forecasting in the fashion industry fashion sales forecasting and the predictive power of online. In: Dynamics in logistics. fourth international conference, LDIC 2014 Bremen, Germany, February 2014 Proceedings, pp 1–4

Boyd DM, Ellison NB (2007) Social network sites: definition, history, and scholarship. J Comput-Mediated Commun 13(1):210–230. doi:10.1111/j.1083-6101.2007.00393.x

Van Eimeren VB (2013) „ Always on “ – Smartphone, Tablet & Co. als neue Taktgeber im Netz. Media Perspektiven 7–8:386–390

Fiddis C (1997) Manufacturer -retailer relationships in the food and drink industry: strategies and tactics in the battle of power. FT Retail & Consumer Publishing, Pearson Professional London

Hines T (2014) Supply chain strategies, structures and relationships. In: Fashion marketing: contemporary issues, 3rd edn. Butterworth-Heinemann, Oxford, pp 27–53

Jackson T (2014) The process of trend development leading to a fashion season. In: Fashion marketing: contemporary issues, pp 169–187)

Jackson T, Shaw D (2001) Fashion buying and merchandising management. Palgrave

Kaplan AM, Haenlein M (2010) Users of the world, unite! The challenges and opportunities of social media. Bus Horiz 53(1):59–68. doi:10.1016/j.bushor.2009.09.003

Khan K, Baharudin B, Khan A, Ullah A (2014) Mining opinion components from unstructured reviews: a review. J King Saud Univ Comput Inf Sci 26(3):258–275. doi:10.1016/j.jksuci.2014.03.009

Peng W, Sun T (2012) Method and system for identifying a key influencer in social media utilizing topic modeling and social diffusion analysis

Uzunoğlu E, Misci Kip S (2014) Brand communication through digital influencers: leveraging blogger engagement. Int J Inf Manage 34(5):592–602. doi:10.1016/j.ijinfomgt.2014.04.007

Weimann G (1994) The influentials: people who influence people. State University of New York Press, Albany

Wong WK (2013) Understanding key decision points in the apparel supply chain. Optimizing decision making in the apparel supply chain using artificial intelligence (AI). Woodhead Publishing Limited, pp 1–12. doi:10.1533/9780857097842.1

Zerfaß A, Bogosyan J (2007) Blogstudie 2007. Inform ationssuche im Internet—Blogs als neues Recherchetoo l

References

Engagement as the Core of Social and Digital Media Strategy in the Fashion Industry

Alicia Gonzalez Miralles, Rinaldo Rinaldi and Romeo Bandinelli

Abstract The main purpose of this paper is to recognize and to underline the role of engagement as the core of a digital and social media strategy. Since many fashion brands haven't exploited completely the potentialities that the continuously evolving digital media can offer, we have reviewed the literature in an effort to structure an approach to define an effective digital media strategy. Furthermore, we examined the Burberry case to identify the axes that addressed brand strategy and the role of engagement in its success. The main contribution of the paper is therefore an engagement-oriented approach to the definition process of a digital strategy that can be implemented by companies who intend to enter the digital business. The current scenario demands a continuously evolving mind-set and organization focused not only on reaching target audiences, but also on retaining them through a customer-centric strategy. The suggested approach might be adopted by companies belonging to various sectors, but customer engagement can be especially useful for visually-oriented sectors like fashion, where consumers are easily reached and already overloaded with content from many sources, and where a captivating and innovative digital marketing strategy can become a critical success factor.

Keywords Customer engagement · Digital strategy · Experience · Contents · Social media monitoring · Omnichannel

1 Introduction

Social networks were originally conceived as virtual spaces to share personal information, pictures, experiences, feelings, and opinions. Joining a social network became viral: millions of users were attracted by the compelling necessity to be "present" and by the satisfaction found through the published contents. Enabling

---

A.G. Miralles (✉) · R. Rinaldi · R. Bandinelli
Department of Industrial Engineering, University of Florence, Florence, Italy
e-mail: alicia.gonzalez@unifi.it

R. Rinaldi and R. Bandinelli (eds.), *Business Models and ICT Technologies for the Fashion Supply Chain*, Lecture Notes in Electrical Engineering 413,
DOI 10.1007/978-3-319-48511-9_23

users to access such masses of information stimulated the need to constantly get and give more, gradually turning these interactions into habits. The intrinsically engaging feature of social and digital media is a strong component of strengthening the importance of engagement as the focus of social- media (SM) strategy.

While the phenomenon was becoming massive, many marketing strategists around the world realized the vast potential of SM. Nonetheless, fashion brands (especially luxury brands) were initially reticent to adopt SM in their marketing strategy due to the contrast between its massiveness and the exclusivity of luxury goods. Recently, this has changed: as a sector in which visual merchandising is a main topic, the Fashion Industry started recognizing social and digital media as a great opportunity to communicate and interact directly with consumers.

Both practitioners and researchers hold that the power of social and digital media can be maximized if used effectively, and supported, for instance, by a well-structured strategy definition, leading to major results in terms of consumer engagement and brand positioning, with the ultimate goal of inducing brand loyalty.

2 Literature Review

2.1 Scientific Background

Brodie et al. (2011) defined customer engagement as a psychological state caused by interactive experiences under diverse contexts, consequently reaching various levels of CE (Gambetti 2012). Brodie described customer engagement as a dynamic, iterative process which creates value within service relationships. He also noted its role in relational concepts, such as loyalty and involvement, which can be antecedents or consequences of nomological networks governing service relationships. Philipp Kotler, an American marketing guru, also remarked on the active role of customers if engaged, stating that, as long as consumers play a key role in the creation of value by product and service co-creation, marketers should focus on the development of meaningful connections with costumers. (Achrol 2012).

Mollen and Wilson (2010) defined online engagement as a cognitive and affective commitment to an active relationship with the brand encapsulated in digital tools which communicate brand value, underlining its dimensions as dynamic and sustained cognitive processing on the satisfaction of instrumental value (utility and relevance) and experiential value (emotions). In 2012, Sashi underlined that, with effective execution, SM can be a valuable communication instrument to enhance brand consumer engagement. This was reinforced by Erdogmus and Cicek (2012), who studied the impact of SM marketing on brand loyalty and concluded that effective campaigns on social media are “the most significant drivers of brand loyalty”. They also identified content relevancy, popularity among friends, and omnichannel as other main drivers (Nadeem 2012; Thipathi 2014).

Engaging customers is the result of a cultivated, bidirectional relation, and, similar to all human relationships, this is not an immediately reachable result. Sashi (2012) defined a seven-stage cycle to describe the process that aims to shape customer engagement: connection, interaction, satisfaction, retention, commitment, advocacy, and engagement. Sashi states that customer engagement is the state where customers feel high levels of emotional attachment and have logical reasons for loyalty, which can result in behavior exhibiting brand advocacy and unconditional loyalty. Consequently, fashion brands should "distribute" their efforts to guide their customers through the entire cycle, while avoiding an early interruption of the link. For instance, a brand might have outdated web content because it focused only on the connection-cycle stage and ignored the channel feeding, thus weakening the retention stage and probably leading to the cycle break point. As stated by Erdogmus and Cicek (2012), providing relevant and updated content is one of the most crucial strategies for managing a brand on SM successfully (Kumar 2012).

As the core of social and digital media strategy, the engagement should be one of the main criteria for evaluating digital/social campaign results. In fact, Hofmann and Fodor (2010) stated that effective social-media measurement should start by converting the traditional ROI approach by calculating the returns in terms of customer social-media investments motivated and driven by brand-versus-consumer motivation to use social media.

2.2 Industrial Background

Some interesting trends regarding the adoption of a SM strategy were presented on the latest Digital Competitive Map (Solca 2016), a report edited by Exane BNP Paribas, which benchmarks the performance of luxury brands on digital media. The analysis is conducted along two axes: "Strategic Reach" (SR) and "Digital Customer Experience Proficiency" (DCEP), both with their own evaluation criteria grouped by categories. It's notable that a separated category for Customer Engagement is present on both axes.

The general results showed that the digital competitive map score grew by 9% in comparison to the previous seasonal readings, while two-thirds of brands in the panel showed significant improvements, meaning that luxury brands are investing in order to increase their performance on digital media. The best performer was Burberry, which stood out and maintained its overall leadership. Fendi, Loro, Piana, and Dior showed the largest improvements on the new Digital Competitive Map.

It was emphasized that there is still no consensus in terms of digital execution priorities because different brands led on different performance criteria: Burberry and Loro and Piana lead in key product presentation criteria. Burberry excelled in personal services–in particular for style advisor, conducted by phone, email and online chat. Dior also excelled in style advisor. Coach led on the all-important cross-channel-services criteria, while Burberry and Fendi were leaders in customer engagement through email proficiency. (These criteria belong to the Digital

Costumer Experience axis). Other noticeable implementations on Digital Costumer Experience were in online order in store, PayPal (Burberry), store search, free delivery (Tory Burch), extra effort on style advisor, collect in store, product exchange in store (Cucinelli), you may also like, match with, ask for visitor localization on desktop/mobile, and web landing pages on emails (Moncler).

Also L2, a member-based business intelligence firm that benchmarks the digital performance of brands, recognizes Burberry as the best in its Fashion Digital Index, which measures the initiatives of 83 luxury brands on websites, e-commerce portals, digital marketing, and social-media presence. Burberry outperformed Kate Spade, Ralph Lauren, Louis Vuitton, and 79 other brands with its activities on Periscope and Snapchat and with its prolific mobile-platform improvements which translated into notable increases in social engagement and brand visibility and a tripling of online sales. These results were published by other digital media like Pambianco, Digiday, fashionmagazine, http://wwd.com, etc.

In the next chapter, Burberry's digital strategy will be studied to extract some best practices from the top digital luxury brand of 2015.

2.2.1 Burberry Case

Back in 2006, Ahrendts and Bailey, Burberry's CEO and CCO (now CEO and CCO), clearly stated that "the vision was to be the first company who is fully digital" looking forward "to build a social enterprise". Ahrendts also mentioned to the Harvard Business Review that the strategy was completely centered on the brand and that, to purify the brand message, they would focus also on digital.

After that announcement, in 2009, came the launch of "Art of the Trench", a microsite that runs user-generated content of people wearing Burberry's signature trench coat, as the first act of the digital transformation (It was upgraded and expanded at the end of 2014). In the same year, Burberry launched its Facebook page and live streamed the Spring/Summer 2010 fashion show, thus overcoming the paradigm of exclusively live runway action.

Then the Burberry Acoustic project showcased "young British bands that Burberry believes in", and, since 2010, consumers can watch videos of the bands performing music and wearing clothes from Burberry's collections.

Burberry has been among the first brands to test new social media channels and developments: In 2013, they posted a video on Instagram while announcing the launch of Instagram for video, and they tested instant orders via text messaging on WeChat in China; In 2014, just after Twitter's buy buttons rolled out, Burberry incorporated the "buy now" button on the Twitter profile; In April 2015, they jumped on Snapchat and Periscope, a Twitter app for live streaming videos which was launched in March 2015, during the live-streamed Spring/Summer 2016 fashion show, and shared shots of the entire collection before it premiered on the runway, driving excitement and engagement among the public to the tune of 100 million impressions. In July 2015, the new service Apple Music streaming was activated, and three months later Burberry became the first brand to get an Apple

Music channel. At the end of September 2015, Burberry and Kakao, Korea's largest social platform with 190 million followers, announced a global partnership that enables simultaneous activities across Kakao Talk, Kakao TV, and Kakao Giftshop and offers Korean audiences direct access to its runway shows and campaigns. It closed the year 2015 with the first 3D interactive marketing campaign in partnership with DreamWorks Animation, enabling five participants, in turn, to pick and personalize a scarf using their smartphones, then to display their 'creations' on the famous Curve screen of Piccadilly Circus (London), and finally to buy it online.

Burberry has partnered with Google twice: the first time in June 2013 for "Burberry Kisses" that used "lip detection technology" to enable users to send digital kisses to loved ones; the second time for "Burberry Booth" on which holiday shoppers could create a 15-second edited version of the Festive campaign in which they appear alongside such celebs and the clip is instantly sent to the user, to motivate the replication of the in-store experience.

Many other in-store experiences have been implemented: in-store digital events; the Burberry Retail Theatre was the first ever live simultaneous virtual trunk shown in stores globally, broadcasting multifaceted content all over the world directly to the stores and in which the audience was able to interact with the collection through iPads and to order online.

Burberry dispatched a Digitally Enhanced Flagship Store equipped with full-length screens around the store which can easily be converted into mirrors and RFID-embedded clothes to enable shoppers at Burberry's Regent Street flagship in London to get specific contents like information about the materials used in a bag or a catwalk videos, when approaching one of the screens in a fitting room.

The brand became an example of digital-media strategy, proving that engaging consumers is possible by presenting the broader culture of the brand with social media contents, even if the company's core business is fashion. Bailey declared that the goal was to bring customers closer to the brand by the mind-set "Mind share if you want market share".

WaveMetrix, a social media monitor, released some results in 2010 about how consumers "thank Burberry" for sharing the "quality music" on Burberry Acoustic and "providing us with great new artists". It also reported that consumers "love" the "amazing" fashion collections and "can't wait" to see the designs in stores. These results showed that social-media content encouraged consumers to discuss the Burberry brand and consider purchase.

Burberry structured a robust, internal digital team that creates the content shared on platforms and on mobile, thus reinforcing the consistency message on digital channels. The transformation encompassed three new departments Social Media, Mobile, and Insight and Analytics; this brought IT to the front because it was realized that a Marketing Department couldn't execute without a strong technology partner. In addition, a special group was dedicated to wearable tech and innovation (the "What If Group").

In the brand's Strategic Report 14/15, Bailey affirmed that highlights included: a 9% increase in comparable store sales, reflecting Burberry's strategic focus on the retail channel; the sustained outperformance of the digital business; and double-digit growth in the Americas and EMEIA regions, each underlining satisfactory results for a digital approach. L2, according to Reuters, reported that the whole of the luxury goods market grew only 1–2% in September 2015, while Burberry's revenue was up 11%.

Furthermore, Burberry's chairman Sir John Peace announced, looking ahead to 2015/2016, that "key investments will include stores in flagship markets, technology and continued digital enhancements".

The four key themes of Burberry's strategic agenda, Peace stated, will be "Brand first, Customer-centric, Famous for product, Productive and responsible". Interestingly, the first and second themes included, as Burberry's long-term priorities, the global engagement driven by innovative creative content and experiences, supported by digital, social, and traditional media in addition to online and in-store innovations, which work together to create a seamless experience in order to be sector-leading in understanding, engaging, and serving its customers, both online and offline.

3 Methodology

This study consisted of two parts. In the first one, already-existing literature reviews on CE have been examined in order to gain an understanding of its relationship with digital and social media and consequently its role in digital strategy. Especially important was the engagement cycle proposed by Sashi (2012) and reinforced by his statement about the importance of engagement to increase customers' emotional attachment which would potentially turn into brand advocacy and unconditional loyalty behaviors.

In the second part, after identification of the best performer in digital strategy in the luxury-fashion industry, we examined the Burberry case to recognize the axes that comprised a brand's effective digital strategy, and we analyzed how engagement played a role in this success.

Burberry's best practices showed four common significant patterns/features: identity (same brand's voice across all platforms); omnichannel (use all channels to reach consumers); integration (of the different social and digital platforms); and continuous innovation (use the latest tech innovations to surprise and consequently engage costumers).

Particular aspects of Burberry's best practices include: delineating customer-centric objectives and consequently campaigns; innovating in order to captivate and engage customers–through all channels as pioneers for some functionalities and by collaborations with many different partners; adopting a mind-share approach to overcome the paradigm a the exclusivity of luxury fashion; implementing in-store improvements to enhance the a customer experience; and having a dedicated team for innovation and content creation.

Finally, after this analysis and taking into account some of the identified best practices, one can endorse an engagement-oriented approach devised to support a brand while defining a customer-centric digital strategy.

4 The Proposed Model

In this chapter, we propose an engagement-based approach to define the processes of the digital strategy for a fashion brand. Initially, we describe some critical issues that can delay or block the workflow when switching to digital will be mentioned, and then we consider some of the best practices identified in the previous chapter.

The first issue a fashion brand faces, when considering an investment in social or digital media projects, is its lack of expertise in the field, due to its different core business. Furthermore, the initial approach usually anticipates calculation of a ROI from social media, for which there isn't yet an accurate, valid model.

Another difficulty brands may encounter is not having the appropriate organizational structure to support a multi-dimensional project because that constitutes implementation of a social-media strategy. In fact, a dedicated (internal or external) team, for content management and/or generation with IT skills to handle the different platforms, could be necessary. Conversion to digital and social media is not (only) about reaching consumers through all the channels but, more importantly, about engaging them with the brand and its products which, if successfully driven, would take consumers to a spontaneous replication of brand's message. So, a constant feeding of the channels is required, which translates in greater effort and resources for the brand.

Moreover, having a dedicated team will make it easier to maintain a unique brand voice while being omnichannel and platform integrated, which is crucial for the brand's identity protection.

To achieve significant results, the brand should be ready to change its mind-set in order to proceed with a structured-strategy definition, while considering that the adopted approaches can be infinite thanks to the variety of channels and possibilities to be present 'online'. This means to be open-minded to overcome some of the previously mentioned issues through a well-structured strategy definition, keeping in mind that building lasting relationships is never an immediate result.

The suggested strategy definition, focused on engagement, starts with the iteration and assessment of these three closely-related steps:

1. *Define the engagement-oriented goals*. State the engagement-oriented goals based on the actual state of the brand in terms of awareness and engagement by using Social Media KPI in order to quantitatively evaluate the initial state, then to set target KPI values and, finally, to measure the concrete results. There must be a definition of specific, measurable, attainable, relevant, and time–bound, customer-centric objectives to drive the entire process to result in successful engagement.

2. *Choose the target audience and channels*. The choice of the target audience and the channels is the primary determinant for both the type and quantity of the contents to be generated and managed, and for the identification of specific indicators to measure reach, customer engagement, and replication.
3. *Define the resources*. Establishing resource availability in order to evaluate if the activities might be executed through a collaboration between the existing areas of the company (IT and marketing), or if it is necessary to define a dedicated team for content management and/or generation.

Once these aspects have been assessed, it is possible to proceed with a content plan and proposal which should be evaluated before its implementation:

4. *Define the content plan*. Content planning which includes posting a calendar and the frequency for related content or campaigns for the chosen channels and subsequent assignment of activities to the allocated resources.

The implementation steps should be:

5. *Define the brand voice*. Establishing a unique brand voice for all the channels and/or adapting internal communication guidelines for the contents that will be generated and published to engage customers while protecting the brand's identity.
6. *Generate content*. Generating a variety of attractive content in order to keep audiences engaged.

Finally, it is necessary to evaluate the impact and effectiveness of the implemented actions by measuring the specific indicators for each channel, using SM analytics software, and analyzing the results and trends for strategy optimization for the next iteration.

5 Discussions and Conclusions

The main purpose of this paper has been to recognize and to underline the role of engagement as the core of a digital and social media strategy.

Many fashion brands are still missing the opportunities that the continuously evolving digital media can offer.

The literature was reviewed in an effort to structure an approach for the definition of an effective digital media strategy. Furthermore, the Burberry case was studied to identify the axes that addressed the brand's strategy and the role of engagement in its success. As a result, we proposed an engagement-oriented approach for the process of defining digital strategy to be implemented by companies that intend to enter the digital arena. The current scenario demands a continuously evolving mind-set and structure focused not only on reaching audiences, but also on retaining them through a customer-centric strategy.

The suggested approach might be adopted by companies belonging to any of various sectors but customer engagement can be especially useful for visually-oriented sectors like fashion, where, since consumers are easily reached and to whom content can be delivered, a captivating digital marketing strategy can make a difference.

In conclusion, it can be stated that engagement plays a crucial role in any social-media strategy. Companies should switch to a mind-set that considers direct interactions with costumers as an opportunity to build a lasting relationship which will gradually drive replication and revenue growth, by offering content strategically constructed to be consistent, continuously innovative, diverse, and consistent.

6 Further Developments

A first step in the future action plan of the project is the validation of the proposed model through enactment of several case studies, thus evaluating performance in order to optimize the current work. Other further developments could involve in-depth study of social-media ROI modelling or delineation of a focused analysis of the interactions and on-line interests of a brand audience and its relationship with content management in order to optimize the digital strategy.

Finally, evaluating the performance of some the most recent digital trends in customer experience (micro-moments, programmatic ads, beacons, etc.) could be interesting in order to analyze their effectiveness and to contemplate their adoption in the digital strategy.

References

Achrol RS, Kotler P (2012) Frontiers of the marketing paradigm in the third millennium. J Acad 40:35–52

Brodie RJ, Hollebeek LD, Juric B, Ilic A (2011) Customer engagement: Conceptual domain, fundamental propositions, and implications for research. J Serv Res: JSR 14(3):252–271

Erdogmus IE, Cicek M (2012) The impact of social media marketing on brand loyalty. Soc Behav Sci 58:1353–1360

Gambetti RC, Graffigna G, Biraghi S (2012) The grounded theory approach to consumer-brand engagement: the practitioner's standpoint. Int J Mark Res 54(5):659–687

Hoffman D, Fodor M (2010) Can you measure the ROI of your social media marketing. MIT Sloan Manage Rev 52(1):40–50

Kotler P, Kartajava H et al (2010) Marketing 3.0: from products to customers to the human spirit. Wiley

Kumar V, Mirchandani R (2012) Increasing the ROI of social media marketing. MIT Sloan Manage Rev 54(1):54–61

L2—Digital IQ Index. www.l2inc.com

Mollen A, Wilson H (2010) Engagement, telepresence and interactivity in online consumer experience: Reconciling scholastic and managerial perspectives. J Bus Res 63(9–10):919–925

Nadeem M (2012) Social customer relationship management (SCRM): how connecting social analytics to business analytics enhances customer care and loyalty? Int J Bus Soc Sci 21(3):88–102

Sashi CM (2012) Customer engagement, buyer-seller relationships, and social media. Manag Decis 50(2):253–272

Solca L et al (2016) Digital competitive map, pp 5–50. www.exanebnpparibas-equities.com

Strategic report 2014–15 – Burberry, pp 9–17. http://www.burberryplc.com/documents/ar-1415/burberry_strategic_report_2014-15.pdf

Thipathi MN (2014) Customer satisfaction and engagement—customer retention strategies for brand manager. Vilakshan, XIMB J Manage 11(1):123–134

QR Code and the Wine Sector: What Contents? An Exploratory Research Study on the Wine Industry

Diletta Acuti, Lorenzo Magherini, Valentina Mazzoli, Romeo Bandinelli, Raffaele Donvito, Rinaldo Rinaldi and Gaetano Aiello

Abstract This paper presents an explorative analysis of the contents shared through QR codes as a tool of proximity marketing. The aim of the research is to identify the most common links shared through the QR code technology, in order to investigate the use of this tool from a marketing point of view. Moreover, the research explores the main aspects of the communications related to the contents shared through QR codes. Specifically, the authors focus on hedonic aspects and functional ones. Finally, the research aims at identifying preliminary best practices in the use of QR code technology. Due to the widespread use of this tool in the sector, the research focuses on the wine industry, as one of the most involved with QR code practices. The study adopts a qualitative approach based on a content analysis of 91 wine labels. Considering the importance of this integrated communication, the purpose of this explorative analysis is to deduce implications that enable managers to master the use of this tool, exploiting all its potential.

Keywords Wine · Marketing · QR code · Social media · N-vivo · Winery · Traceability · Content marketing · Tourism · Sustainability

1 Introduction

We often characterize the agro-food industry as one with little value-added and little innovative content. In such a globalized world, this sector contains considerable opportunities for technological and rent upgrading. As confirmed by recent research (Farinelli 2012; Kaplinsky and Fitter 2004; Kaplinsky and Readman 2005), there is

D. Acuti · V. Mazzoli · R. Donvito
Economics and Management Department, University of Pisa, Pisa, Italy

L. Magherini (✉) · R. Bandinelli (✉) · R. Rinaldi · G. Aiello
Industrial Engineering Department, University of Florence, Florence, Italy
e-mail: lorenzo.magherini@unifi.it

R. Bandinelli
e-mail: romeo.bandinelli@unifi.it

R. Rinaldi and R. Bandinelli (eds.), *Business Models and ICT Technologies for the Fashion Supply Chain*, Lecture Notes in Electrical Engineering 413,
DOI 10.1007/978-3-319-48511-9_24

an ongoing process of de-commodification of primary goods, which are increasingly being transformed from everyday staples into high quality products, diversified both in terms of aesthetics and for the changing needs of consumers (e.g., the tendency of some adults to develop vegetarian or vegan dietary habits). Others become processed products by means of the accumulation of greater knowledge and technological progress, which increase the value of content shown as well as the export price per unit. Wine is one of the most interesting cases from a catch-up point of view, because the latecomers in the international market have changed how wine is produced, sold, and consumed and, in doing so, they have challenged the positions held by the incumbents (Giuliani 2011).

Wine is produced, sold, and consumed globally. The quality of the wine, as an agricultural food product, is closely tied to the nature of the original crop (i.e., the grapes). For this reason, wine production is highly dependent, on and influenced by, geographical characteristics.

At the same time, its sales are being developed via marketing strategies that tend to emphasize the characteristics of the production area, the techniques used, the handmade aspects, etc.

In the wine industry this is referred to as terroir.

Terroir is more complex than merely location. It involves climate, soil, tradition, terrain, and even culture by some definitions.

The concept of terroir captures such diversity coupled with history and tradition (Charters 2006), and confers on a certain type of wine a unique competitive advantage over other producers (Wilson 1998; Vaudour 2002; Barham 2003). Wine producing countries, under the regulations of the EU commission, have introduced several schemes and legislation protecting the place of origin of wines and regulating its production in many aspects, ranging from maximum yields per hectare, oenological practices, grape varieties, and the labelling of wine among others. As an example, in Italy the regulation can affect, besides the already mentioned, even the oenological (cuts, practices, etc.), product specification (production area, type of grape, wine-growing practices, and winemaking methods), the permitted indications (DOP, IGP), national endorsements (i.e., DOC for Italy, AOC for France AOC), and authorization to name a wine with the region of origin (i.e., Asti, Porto, Champagne).

The competitive dynamics of the wine market are complex (Orth and Lockshin 2007): the scenario has expanded into a global competition for consumer's taste preferences and on hedonic and emotional aspects, as well as on aspects related to health and sustainability. These innovative dynamics are changing wineries' strategies. Marketing innovation will be important in order to build competitive advantage and achieve growth (Chen 2006). Wineries are recognizing the need to acquire more environmentally efficient technologies on the one hand, and ensure proper management of the product traceability in order to use it as a marketing strategy for the consumers, on the other. RFID technology is one of the developing frontiers in this sector as it will make the wine supply chain completely transparent and advise consumers on the origin and processing of a bottle. The issue of sustainability, both in primary production processes (grape growing) and in its transformation (wine-making), is taking a central role in the operational and strategic choices of wineries, because of the increasing concerns about the environment, ecological consequences, and the efficient use of natural renewable resources as reflected in public opinion and consumers' perceptions (Pilone et al. 2015). This can be used as an important marketing strategy.

There are many studies that highlight the influence that the innovative approach has on the winery's ability to improve the satisfaction level in meeting the consumer's expectations (Santini et al. 2007).

2 Literature Review

Consumer habits are changing as result of new technologies used in day-to-day life (Grewal et al. 2012) and, as a consequence, consumer goods companies are developing new ways of relating to their customers (Peira et al. 2012). Because of the progress and the spread of new technologies, communication strategies have undergone an evident revolution. Mobile diffusion has not only changed the way in which organizations and their brands interact with customers, it has also changed the way business gets done: they are able to reach consumers with digital tools and network with them (Begalli et al. 2012). This is one of the reasons why recent years have witnessed growing interest of marketers in mobile phone as a channel of marketing communication, and it will continue to gain marketers attention even further (Wohlfahrt 2002). The specific characteristics of mobile phones like geo-targeting, ubiquity, immediacy, customization, measurability, and interactivity encourage using mobile phone in hardcore marketing (Bauer et al. 2005; Haghirian et al. 2005). Apart from the above modes of mobile marketing, one very innovative mode that has been catching marketers' attention recently in current digital space, i.e. QR codes (Bamoriya 2014). QR code is an abbreviation for Quick Response Code. Basically it's a 2D code, which, once scanned by a Smartphone—with a scanning application software—connects users to some specific online content on a website, linking to an email address, delivering e-coupons, texting, leading to registration, and so on (Handley 2012; Bisel 2011). Due to the convenience, user-friendliness, multiple fields of application, and customer-driven information provided by mobile tagging, it has become the key technology for mobile surfing (Shiang-Yen et al. 2013). Indeed, product knowledge is a key concept in consumer behavior and consumer decision making, and QR codes have the ability to deliver and share information at the point of need (Raju et al. 1993). QR Codes are a good fit for almost any sector, but for wine they offer a particular benefit (Brabazon et al. 2014). Indeed, the wine sector is challenging the Wine 2.0 era, using the Internet to engage with wine consumers on their terms, in a time and manner of their choosing (Olsen and Hermsmeyer 2008). Tools usually include social-networking sites, blogs, message boards, and other methods that leverage user-generated content. Wine is an experiential good that cannot be fully evaluated until it is consumed, thus in the absence of tastings or prior experience with the wine, the purchase decision is challenging for many wine consumers (Cooper-Martin 1991). Now, thanks to technology, wine purchasing has become part of the information era (Halstead 2013). The importance of using QR codes in the wine sector is due to the peculiar characteristics of this product: many consumers do not simply buy a bottle of wine on the shelf for its particular use; they buy its heritage, quality, territory, production processes, and specific organoleptic features.

Wine is recognized as a lifestyle beverage, and the wine consumer's relationship therewith is based on an acquired, not an innate need, as in the cases of food and water. Wine consumption itself can be regarded as a hedonic experience for some people—an emotional and pleasurable activity aimed at personal enjoyment (whether of a basic wine drinker or a connoisseur) within the context of a myriad of potential social experiences (Bruwer and Alant 2009).

For these reasons, communicating the experience of hedonic goods involves a large portion of the retail marketing strategy, such as managing asymmetric information (Akerlof 1970; Nelson 1970) and resolving this asymmetry through the use of price and advertising (Milgrom and Roberts 1986; Nelson 1974), warranties (Grossman 1981), and expert opinion (Cuellar and Claps 2013; Eliashberg and Shugan 1997; Reinstein and Snyder 2005).

The consumer interest in wine content varies depending on the scanning time: QR codes are not only useful at the point of sale (Brabazon et al. 2014), but also during the consumption. Hence, shared contents should vary depending on the scanning time, considering the trade-off between the hedonic aspects and functional aspects of the product (Ferrarini et al. 2010). Moreover, contents may vary depending on brand awareness of the wine. In the literature, studies have investigated the wine sector from various points of view (Lockshin and Corsi 2012)—retailing, purchasing, tourism (Bruwer and Alant 2009), lifestyle, packaging and labelling, region (Easingwood et al. 2011), sustainability (Forbes et al. 2009), technology (Brabazon et al. 2014), social media (Reyneke et al. 2011)—but there is a gap regarding the contents taxonomy that wine companies communicate to consumers through new digital tools. This exploratory research investigates this issue within the context of the Italian market.

3 Methodology

This exploratory analysis of the uses of QR codes in the wine industry adopts a qualitative approach through a content-analysis methodology (Berelson 1952). Authors scanned 91 QR codes of Italian wine brands purchased from three large stores of grocery retailers located in Tuscany and classified the contents in terms of website homepage or specific website page, the technical file of the wine, or other pages (Yoo and Kim 2014; Shobeiri et al. 2014; Geissler et al. 2006). Hence, authors focused only on website homepages and website technical files of the products with the intent of examining specific hedonic (Alba and Williams 2013; Hirschman and Holbrook 1982) and technical aspects of each category (Sheth et al. 1991).

In particular, the scientific authors codified hedonic aspects, i.e., contents aiming at stimulating an emotional response from consumers, in terms of fun, amusement, fantasy, arousal, sensory stimulation, and enjoyment (Hirschman and Holbrook 1982).

Technical or functional aspects are considered specific features or characteristics of the product, the production, or the winery.

The mind map of the content analysis is specified in Fig. 1. The nodes of classification have been identified top-down (hedonic features and technical feature), and bottom-up (website homepage, website technical file of the product, heritage, quality, territory, certification, sustainability, functional aspects, and organoleptic aspects).

The content analysis has been carried out with N-ivo software (Bazeley and Jackson 2013).

Authors, once codified contents retrieved from QR code links in terms of hedonic and technical aspects and created the respectively nodes, run a word frequency query within the nodes in order to identify the main lemmas used for the website communication.

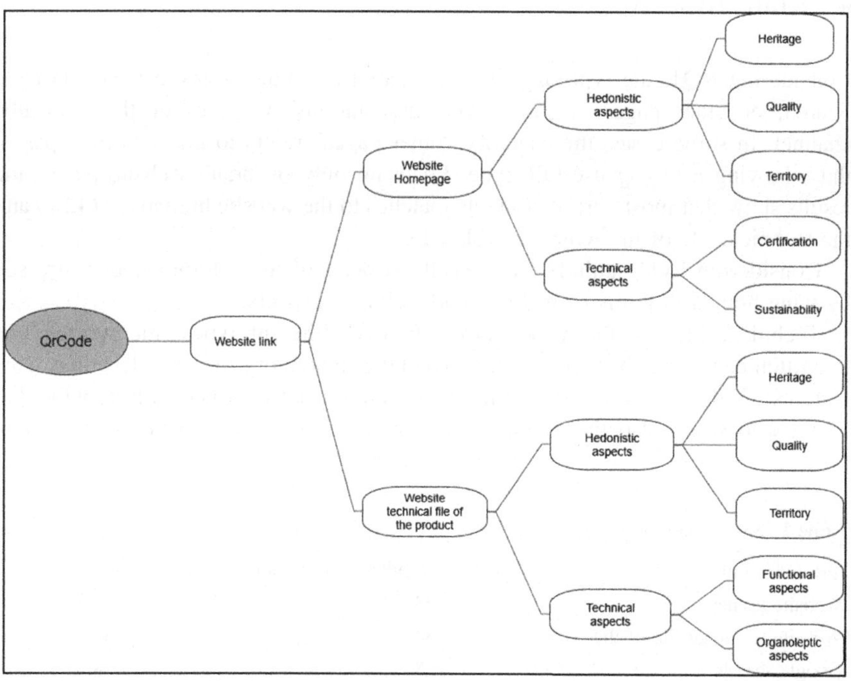

Fig. 1 Concept map of content analysis

The codifying process has been carried out separately by two research who then compared the results with NVivo intercoded reliability function in order to assess the consistency of the codification. The average of the intercoded reliability index within various nodes of codification is 97.0%.

Finally, the authors focused on contents relying on the degree of brand awareness based on official rankings, with the aim to identify a benchmark of best practices within the sample considered. The authors assume there is a relationship between typology of content shared and typology of brand analyzed, in particular, since it was assumed that the consumers of renowned brands are more aware of QR code technology.

Considering the phases previously discussed, the research questions that this study addresses are:

RQ1 What kind of links are shared through QR codes in the wine industry?
RQ2 What aspects of the communication are stressed in the contents shared through QR codes in the wine industry?
RQ3 What are the best practices in terms of QR code use in the wine industry?

4 Main Results

Considering RQ1, the typology of links shared are brand website pages, Google Search, or other pages, such as wine associations' websites or the YouTube channel. In some cases, the category "other pages" refers to not identified pages, thus showing a wrong use QR code. Focusing only on brand website pages, the results show that most part of contents matches to the website homepage (42%) and the technical file of the wine 37 (Table 1).

Considering RQ2, authors analyzed the aspects of the communication stressed by wine brands in terms of hedonic and technical aspects.

Technical files of the wine belong to two different types: the written one (a written page describes wine features) and the spoken one (a video describes wine features). Video generally are not limited to the functional aspects of the wine, but they try to stress also the hedonic and emotional aspects of wine consumption.

Table 1 Sample composition

QR code content	Number of wine labels	%
Website homepage	38	41.8
Website technical file of the wine	34	37.4
Google search	8	8.8
Other pages	9	12.1
Total	91	100.0

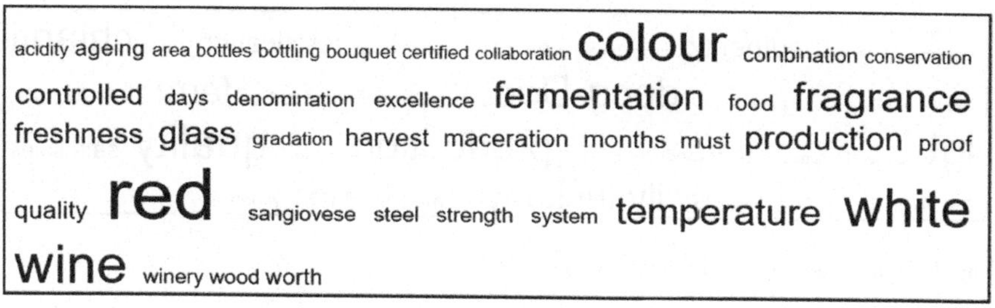

Fig. 2 Technical file: tagcloud within technical node

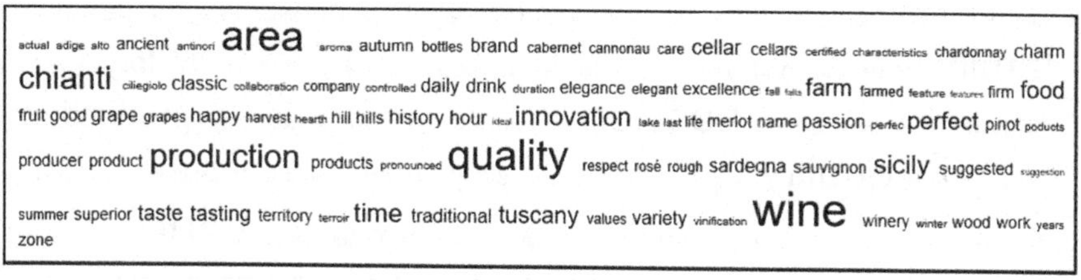

Fig. 3 Technical file: tagcloud within the hedonic node

In addition, written technical files seek to mix the two aspects, but in the largest part of cases, wineries limit the description only to specific functional and organoleptic aspects, such as area of production, terrain, grapes, vinification, color, aroma, and taste. Besides, in some cases, wineries seek to evoke also hedonic aspects, such as the heritage and tradition of the brand. However, functional and organoleptic aspects are more consistent.

In order to derive the main lemmas used in the technical file, Figs. 2 and 3 summarize this information.

Technical associations in technical files of the wine are strongly related to the processes of the production (i.e., wine harvest, maceration, fermentation, ageing) and to the main features of the wine (i.e., color, gradation, acidity).

The most important associations retrieved in the hedonic node of the technical file stress the role of the quality through the innovation of the production used in the territorial area of the winery's cellar.

The same criteria drove the content analysis on website homepages. In this case, hedonic aspects are predominant, focusing on images and text that evoke brand values such as the quality of the production resulting from the strong links to the territory and the brand heritage. The technical aspects identified in this case are certifications (Igt, DOC, DOCG, or Organic Production) and sustainability procedures.

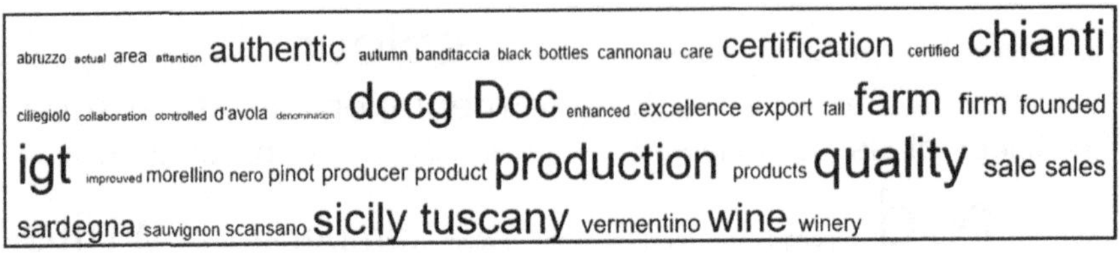

Fig. 4 Homepage: tagcloud within technical node

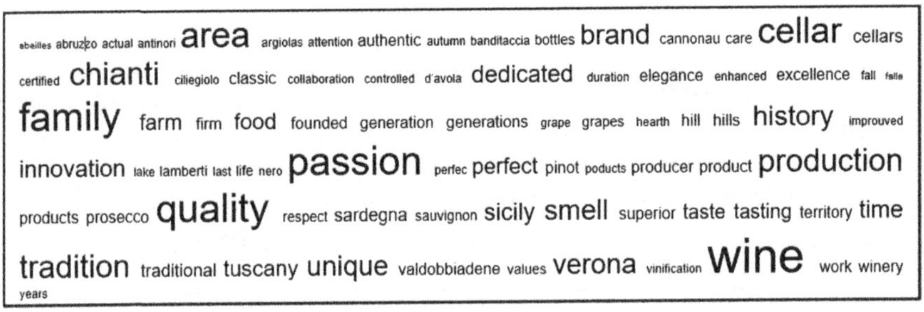

Fig. 5 Homepage: tagcloud within the hedonic node

In order to convey the main lemmas used in the technical file, Figs. 4 and 5 summarize this information.

As we can see from the results shown in Fig. 4, the main associations retrieved from the technical node refer to product certification, area of production, and consortium membership (i.e., Chianti), in order to transmit the quality of the wine.

Lemmas related to the hedonic node of the website homepage refer principally to the heritage of the brand (i.e., family, history, tradition) and the area of production in order to evoke the emotions of consumers.

Finally, in order to answer RQ3, the authors analyzed the results considering only the wineries of the sample included in the official wine rankings in order to find out which type of contents are shared by the most important players in the wine market. Eleven of the 91 wineries were included in the analysis. Seven of out eleven shared the technical files of the product through QR codes, trying to mix technical aspects and hedonic ones, via various media such as video, text, and images.

Moreover, four out of the eleven 11 most important wineries use QR codes to share their website link. Predominately, the websites feature hedonic aspects in terms of the elements previously discussed.

Results are summarized in Fig. 6.

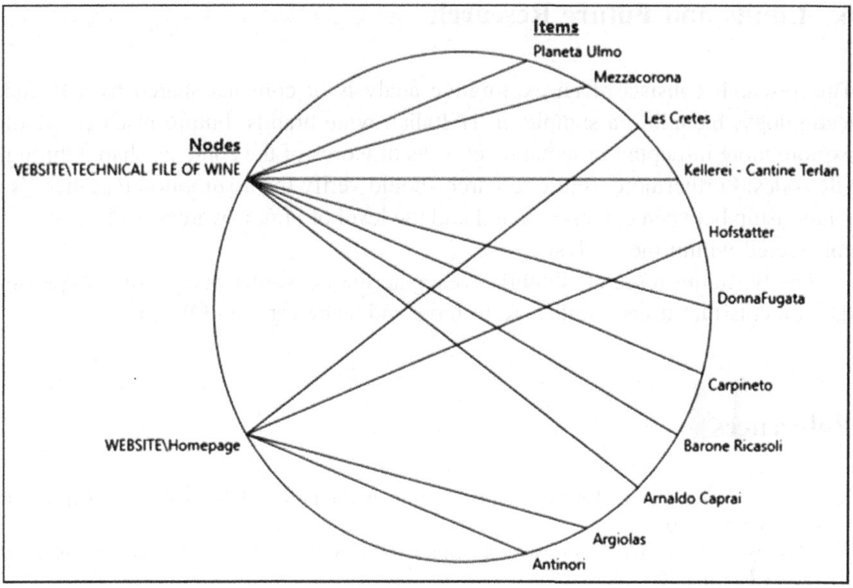

Fig. 6 Homepage: tagcloud within the hedonic node

5 Conclusions

While wineries use QR codes in various ways, brand website pages are the most frequent link shared. Within brand-website pages, technical files of the product and the website homepage are the most common links used. The main aspects of the communication observed have been classified in terms of hedonic and functional contents, where the former refers to the pleasure and the amusement of the consumer and the latter refers to technical features of the product. On website homepages, hedonic aspects prevail, whereas, in the wine technical files, functional aspects are more abundant.

Based on the benchmark from the eleven big players included in the sample, technical files of the product seem to be the most effective content, when considering the practices of the market's big players. However, this information should be sustained by other information more related to hedonic aspects. Indeed, consumers may use QR code to uncover additional information than that normally shared on the traditional label. Therefore, wineries should find a balance between functional and organoleptic aspects, useful for demanding and prepared consumers, and hedonic aspects that may be determinant for other consumers that need to be seduced by such contents. From this perspective, wine is not only a beverage, but also a lifestyle product, such as fashion. Therefore communication should reflect this product condition, transmitting both technical information, very important especially for demanding consumers, but also hedonic information able to affect consumer preferences through emotion.

6 Limits and Future Research

This research consists of an explorative analysis of contents shared by QR code technology, including a sample of 91 Italian wine brands. Future research should explore more in-depth the semantic choices of words of the contents shared through QR codes. Furthermore, future research should verify the assumption that there is a relationship between contents shared and the level of brand awareness of the brands considered within the analysis.

Finally, future research should investigate the consumer perspective, exploring the contents that users would like to have and share through QR codes.

References

Akerlof GA (1970) The market for 'lemons': quality uncertainty and the market mechanism. Q J Econ 84:488–500

Alba JW, Williams EF (2013) Pleasure principles: a review of research on hedonic consumption. J Cons Psychol 23(1):2–18

Bamoriya H (2014) Cross-cultural exploration of consumers' beliefs and behavioral intentions towards QR codes in marketing: an experimental study in India and USA. Acta Universitatis Danubius. Œconomica 10(4)

Barham E (2003) Translating terroir: the global challenge of French AOC labelling. J Rural Stud 19(1):127–138

Bauer HH, Reichardt T, Barnes SJ, Neumann N (2005) Driving consumer acceptance of mobile marketing: a theoretical framework and empirical study. J Electron Commerce Res

Bazeley P, Jackson K (2013) Qualitative data analysis with N-vivo. Sage Publications Ltd

Begalli D, Capitello R, Codurri S (2012). Social media and web marketing strategies: an exploratory study in the Italian wine sector. Sci Papers 21

Berelson B (1952) Content analysis in communication research. The Free Press, New York

Bisel J (2011) Mobile marketing and QR codes. Hudson Valley Bus J September Issue:21

Brabazon T, Winter M, Gandy B (2014) QR codes and wine media. In: Digital wine. Springer, Singapore, pp 55–85

Bruwer J, Alant K (2009) The hedonic nature of wine tourism consumption: an experiential view. Int J Wine Bus Res 21(3):235–257

Charters S (2006) Aesthetic products and aesthetic consumption: a review. Consump Mark Cult 9 (3):235–255

Chen Y (2006) Marketing innovation. J Econ Manage Strategy 15(1):101–123

Cooper-Martin E (1991) Consumers and movies: some findings on experiential products. Adv Consum Res 18:372–378

Cuellar SS, Claps M (2013) Differential effects of brand, ratings and region on willingness to pay: a hedonic price approach. J Wine Res 24(2):138–155

Easingwood C, Lockshin L, Spawton A (2011) The drivers of wine regionality. J Wine Res 22:19–33

Eliashberg J, Shugan SM (1997) Film critics: influencers or predictors? J Mark 61:68–78. doi:10.2307/1251831

Farinelli F (2012) Natural resources, innovation and export growth: the wine industry in Chile and Argentina, Maastricht University

Ferrarini R, Carbognin C, Casarotti EM, Nicolis E, Nencini A, Meneghini AM (2010) The emotional response to wine consumption. Food Qual Prefer 21(7):720–725

Forbes SL, Cohen D, Cullen R, Wratten S, Fountain J (2009) Consumer attitudes regarding environmentally sustainable wine: an exploratory study of the New Zealand marketplace. J Clean Prod 17(13):1195–1199

Geissler GL, Zinkhan GM, Watson RT (2006) The influence of home page complexity on consumer attention, attitudes, and purchase intent. J Advert 35(2):69–80

Giuliani E (2011) Role of technological gatekeepers in the growth of industrial clusters: evidence from Chile. Region Stud Taylor & Francis 45(10):1329–1348

Grewal D, Roggeveen A, Compeau L, Levy M (2012) Retail value-based pricing strategies: newtimes, newtechnologies newconsumers. J Retailing 88(1):1–6

Grossman SJ (1981) The informational role of warranties and private disclosure about product quality. J Law Econ 24:461–489. doi:10.1086/466995

Haghirian P, Madlberger M, Tanuskova A (2005) Increasing advertising value of mobile marketing—an empirical study of Antecedents. In: Proceedings of the 38th Hawaii international conference on system sciences

Halstead L (2013) Consumer trends in the wine industry for 2013. WineViticult J. www.winebiz.com.au

Handley A (2012) Cracking the mobile code. Entrepreneur, pp 56–57

Hirschman EC, Holbrook MB (1982) Hedonic consumption: emerging concepts, methods and propositions. J Mark:92–101

Kaplinsky R, Fitter R (2004) Technology and globalisation: who gains when commodities are de-commodified? Int J Technol Global 1(1):2004

Kaplinsky R, Readman J (2005) Globalization and upgrading: what can (and cannot) be learnt from international trade statistics in the wood furniture sector? ICC 14(4):679–703

Lockshin L, Corsi AM (2012) Consumer behaviour for wine 2.0: A review since 2003 and future directions. Wine Econ Policy 1(1):2–23

Milgrom P, Roberts J (1986) Price and advertising signals of product quality. J Polit Econ 94:796–821

Nelson P (1970) Information and consumer behavior. J Polit Econ 78:311–329. doi:10.1086/259630

Nelson P (1974) Advertising as information. J Polit Econ 82:729–754. doi:10.1086/260231

Olsen J, Hermsmeyer J (2008) Direct wine sales and wine 2.0. In: l. thach and t.matz (eds) Wine: a global business, 2nd edn. Miranda Press, New York

Orth UR, Lockshin L (2007) The global wine business as a research field. Int J Wine Bus Res 19 (1):5–13

Peira G, Bonadonna A, Arnoldi S (2012) Smart tags in the wine sector as a tool for extended packaging: the perception of wine-producing companies in the province of Turin. In: Conference proceeding

Pilone V, De Lucia C, Del Nobile M, Contò F (2015) Policy developments of consumer's acceptance of traditional products innovation: The case of environmental sustainability and shelf life extension of a PGI Italian cheese. Trends Food Sci Technol 41(1):83–94

Raju PS, Lonial S, Mangold W (1993) Differential effects of subjective knowledge, objective knowledge and usage experience on decision making: an exploratory investigation. J Consum Res 15:253–264

Reinstein DA, Snyder CM (2005) The influence of expert reviews on consumer demand for experience goods: a case study of movie critics. J Ind Econ 53:27–51

Reyneke M, Pitt L, Berthon PR (2011) Luxury wine brand visibility in social media: an exploratory study. Int J Wine Bus Res 23(1):21–35

Santini C, Cavicchi A, Rocchi B (2007) Italian wineries and strategic options: the role of premium bag in box. Int J Wine Bus Res 19(3):216–230

Sheth JN, Newman BI, Gross BL (1991) Why we buy what we buy: a theory of consumption values. J Bus Res 22(2):159–170

Shiang-Yen T, Foo LY, Idrus R (2013) Application of quick response (QR) codes in mobile tagging system for retrieving information about genetically modified food. Adv Data Netw Commun Comput. ISBN: 978-960-474-245; ISSN: 1792, 6157, 1417-1426

Shobeiri S, Mazaheri E, Laroche M (2014) Improving customer website involvement through experiential marketing. Serv Ind J 34(11):885–900

Vaudour E (2002) The quality of grapes and wine in relation to geography: notions of terroir at various scales. J Wine Res 13(2):2002

Wilson J (1998) Terroir: the role of geology, climate and culture in the making of French wines. University of California Press, Octopus Publishing Group

Wohlfahrt J (2002) Wireless advertising, mobile commerce: Grundlagen. Erfolgsfaktoren, Günther Silbere, Wiesbaden, Press, Geschäftsmodelle

Yoo J, Kim M (2014) The effects of home page design on consumer responses: moderating role of centrality of visual product aesthetics. Comput Hum Behav 38:240–247

The manufacturer's authorised representative in the EU is Springer Nature Customer Service Centre GmbH, Europaplatz 3, 69115 Heidelberg, Germany. If you have any concerns regarding our products, please contact ProductSafety@springernature.com

Printed and bound by CPI Group (UK) Ltd, Croydon, CR0 4YY
30/09/2026
02225570-0008